Renewable Energy for Mitigating Climate Change

Renewable Energy for Mitigating Climate Change

Edited by
Jacqueline A. Stagner and David S-K. Ting

CRC Press
Taylor & Francis Group
Boca Raton London New York

CRC Press is an imprint of the
Taylor & Francis Group, an **informa** business

First edition published 2022
by CRC Press
6000 Broken Sound Parkway NW, Suite 300, Boca Raton, FL 33487-2742

and by CRC Press
2 Park Square, Milton Park, Abingdon, Oxon OX14 4RN

Library of Congress Cataloging-in-Publication Data
Names: Stagner, Jacqueline A., editor. | Ting, David S-K., editor.
Title: Renewable energy for mitigating climate change / edited by
Jacqueline A. Stagner and David S-K. Ting.
Description: First edition. | Boca Raton, FL : CRC Press, [2022] |
Includes bibliographical references and index.
Identifiers: LCCN 2021027696 (print) | LCCN 2021027697 (ebook) |
ISBN 9780367758110 (hbk) | ISBN 9781032146072 (pbk) |
ISBN 9781003240129 (ebk)
Subjects: LCSH: Renewable energy sources. | Climate change mitigation.
Classification: LCC TJ808 .R4245 2022 (print) |
LCC TJ808 (ebook) | DDC 333.79/4–dc23
LC record available at https://lccn.loc.gov/2021027696
LC ebook record available at https://lccn.loc.gov/2021027697

ISBN: 978-0-367-75811-0 (hbk)
ISBN: 978-1-032-14607-2 (pbk)
ISBN: 978-1-003-24012-9 (ebk)

DOI: 10.1201/9781003240129

Typeset in Times LT Std
by Newgen Publishing UK

Contents

Preface

This book aims at disseminating the latest progress concerning the marriage between renewable energy and climate change. Based on the state-of-the-art developments, the way to further capitalize on renewable energy for mitigating climate change is unraveled. The volume confronts climate change from multiple fronts, conveyed by the experts in various domains around the globe. Both high-level, more general disclosures and in-depth, technological advancements are judiciously presented. Potent approaches and promising out-of-the-box ideas are particularly highlighted. The interests of all stakeholders, especially future generations, form the thread connecting all the chapters together into a powerful weapon to fight climate change.

Yamsrual and Potipituk begin this volume by reminding us that the global challenge can be effectively overcome via local actions in Chapter 1. They emphasize creating an eco-industrial estate targeting low-carbon property with localized self-generating renewable energy. Floating solar is particularly attractive according to their case study performed for Thailand. Talking about the promise of solar, Hossain presents a calculative mechanism to estimate the global solar energy to present a clean and renewable energy system for the world in Chapter 2. The calculated numbers assert that the available solar energy is over ten thousand times the current energy usage. It's time for us to start tapping more into what the sun is lavishing upon us—free and clean energy. The good sun provides more than direct solar energy but also another green energy, wind, in abundance. Chapter 3 features energy storage as a practical way to increase wind energy integration. Womble and Xydis take the United States as their case study. They conclude that, via appropriate energy storage, we can eliminate curtailment during the time supply exceeds demand. In the US setting, this curtailment elimination alone can supply enough energy for up to an average of 3 million households. There is more on renewable energy that we can take advantage of; improved management with real monetary incentives can substantially enhance the current clean energy status. Gökgöz and Erdoğan make use of a portfolio financial optimization technique for geothermal power plants in the Turkish electricity market in Chapter 4. Their empirical study helps energy market decision makers on how to establish a viable decision support system.

When discussing sustainability, it is inevitable for us to revisit the good old ways. In Chapter 5, Balo and Polat take us back to the beautiful natural stones that dominated traditional buildings. They single out an authentic restaurant in Harput region for their illustration. Certain traditional building materials can simultaneously retain the beautiful regional culture, bioclimatic comfort, and significant reduction in energy consumption. Recycling of building waste is the focus of Chapter 6 by Halabi. Integrating wastes, including plastic, into concrete is in line with the Sustainable Development Goal 9 set by the United Nations. A proper balance of the different recycled materials with base concrete into a functional building material is a worthwhile art in architecture. One emerging technology along this line is 3D printing, where a wealth of studies has emerged in recent years. Abdallah, Afsar, Estévez, and Popov unravel the added challenge in bringing 3D printing up to the architectural scale in Chapter 7.

They experiment with cyber-physical manufacturing mode for remote 3D printing of various bio-composite materials in their case study. One highlight is the synthesizing of three different material compositions and calibrating them remotely in the 3D process while maintaining real-time calibration-feedback networking and communication between the designer and the manufacturer.

Castillo, Gennett, Estévez, and Abdallah affirm that seclusion is bad in Chapter 8. They propose a tri-symbiosis between humans, pigeons, and algae for supporting environmental, economic, and urban design sustainability. Namely, algae as food for pigeons and humans, relieving economic stress on grains, and enhancing human–pigeon interaction. The key is to realize mass cultivation of micro-algae, *Chlorella* spp, in open pond photobioreactors in a cost-effective and eco-friendly manner. What about the water? With agriculture utilizing roughly 70 percent of all water abstraction worldwide, Expósito and Velasco expound on the challenge in implementing the European Union Water Framework Directive for transboundary waters in Chapter 9. They use the data envelopment analysis to assess the capacity of selected European river basins to minimize agricultural pressures on water resources.

This volume concludes with the somewhat sensitive topic of nuclear energy. As explained by Schuelke-Leech and Leech in Chapter 10, nuclear power is akin to renewables in its ability to provide clean energy. Great strides have been made in recent years to overcome the issues associated with nuclear energy. And, as Bill Gates advocates for it, maybe we need to take another serious look at nuclear energy.

Acknowledgments

We are indebted to the wonderful men and women who toiled to make this timely volume a reality. At the forefront are the experts in the respective fields who compiled the ten chapters. Thanks to the anonymous reviewers who sharpened the manuscripts. The editors are getting used to the excellent Taylor & Francis team, Joseph Clements, Lisa Wilford, and others working behind the scenes. It would be remiss of us if we do not acknowledge the grace from above that sustained the project from the onset to finish.

Editor Biographies

Jacqueline A. Stagner is the Undergraduate Programs Coordinator of Engineering at the University of Windsor. She has a PhD in materials science and engineering, a masters in business administration, and a bachelor's degree in mechanical engineering. As an adjunct graduate faculty member in the Department of Mechanical, Automotive and Materials Engineering, she co-advises students in the sustainability and renewable energy areas, in the Turbulence & Energy Laboratory. To date, she has edited four volumes.

David S-K. Ting is the founder of the Turbulence & Energy Laboratory, University of Windsor. As a professor in the Department of Mechanical, Automotive and Materials Engineering, he supervises students on a wide range of research projects in the thermofluids, flow turbulence, energy conversion and conservation, and renewable energy areas. Professor Ting has supervised over 85 graduate students and co-authored more than 150 journal papers.

Contributors

Yomna K. Abdallah
iBAG, UIC Barcelona (Institute for
Biodigital Architecture & Genetics)
Universitat Internacional de Catalunya

Secil Afsar
iBAG, UIC Barcelona (Institute
for Biodigital Architecture &
Genetics)
Universitat Internacional de
Catalunya

F. Balo
Department of Industrial Engineering
Firat University

Javier G. Ca12stillo
Master in Biodigital Architecture
ESARQ-UIC Barcelona School of
Architecture
Universitat Internacional de
Catalunya

Ahmet Yıldırım Erdoğan
Project Manager, Turksat Company
Ankara, Turkey

Alberto T. Estévez
ESARQ UIC Barcelona School of
Architecture
Institute of Biodigital Architecture and
Genetics
Universitat Internacional de Catalunya

A. Expósito
University of Malaga, Spain

Andrew Gennett
Master in Biodigital Architecture
ESARQ-UIC Barcelona School of
Architecture

Universitat Internacional de
Catalunya

Fazıl Gökgöz
Professor, Department of Management,
Faculty of Political Sciences
Ankara University
Ankara, Turkey

Marwan Halabi
Associate Professor, Faculty of Architecture,
Design and Built Environment
Beirut Arab University

Md. Faruque Hossain
College of Architecture and
Construction Management
Kennesaw State University,
Marietta, GA

Timothy C. Leech
Independent scholar

H. Polat
Department of Architecture
Firat University

Oleg Popov
3DPlodder 3D printing, Ukraine

Chantamon Potipituk
Assistant Professor in Architecture
Faculty of Architecture and Design
Rajamangala University of Technology
Rattanakosin
Salaya, Phutthamonthon District,
Nakhon Pathom, Thailand

Beth-Anne Schuelke-Leech
University of Windsor

F. Velasco
University of Seville, Spain

Marissa Schmauch Womble
Energy Policy and Climate Program
Krieger School of Arts & Sciences
Johns Hopkins University, Baltimore,
    Maryland

George Xydis
Department of Business Development
    and Technology
Aarhus University
Herning, Denmark

Supattra Yamsrual
Environmental and Energy Division
Environmental Department
Industrial Estate Authority of Thailand
Bangkok, Thailand

1 Furthering Renewable Energy for Climate Change Mitigation

Case Study on the Implementation in Eco-Industrial Estates in Thailand

Supattra Yamsrual and Chantamon Potipituk

1.1 INTRODUCTION

Over the last few decades, climate change has become a hot topic for researchers and public due to its massive impact. The impacts of climate change are immensely felt on biodiversity, ecosystem, health, livelihood, food security, water supply, human security, and economic growth, especially when the global warming exceeds 1.5°C than at present (IPCC, 2018a). Thus, nations are taking steps to encounter them, and the United Nations emphasizes this issue with the 13th goal of Sustainable Development Goals (SDGs), which calls for an urgent action to combat climate change and its impacts. The two measures for tacking climate change—climate change mitigation and climate change adaption—have been vastly implemented in many countrywide strategies. Mitigation, therefore, targets the causes of climate change, while adaptation addresses its impacts (UNESCO, 2019). Mitigation measures are the actions or activities that are taken to reduce the sources or enhance the sinks of greenhouse gases (IPCC, 2014). Xinsheng et al. (2008) found that the solutions of climate change solution placed more emphasis on mitigation strategies than adaptation behavior; however, the effective climate change solution requires the combination of mitigation

DOI: 10.1201/9781003240129-1

1

and adaptation, which involves reducing vulnerability to the effects of climate change (Nyong et al., 2007; Vijaya Venkataraman et al., 2012).

There are some of the mitigation measures that can be taken to lessen greenhouse gas emissions such as practicing energy efficiency, increasing the use of renewable energy, and efficient means of transport. Maximizing the use of renewable energy would be essential for climate change mitigation. The IPCC reported that the renewable energy will be targeted to supply 70–85 percent of electricity in 2050 (IPCC, 2018b). Renewable energy has to be utilized in all the sectors, including the industrial sector, which contributes 21 percent of the total greenhouse gases emitted (ranked third after energy sector and agriculture and land use sector) (IPCC, 2014). Therefore the use of renewable energy as a replacement for current sources of energy is considered as an effective mitigation strategy in the industrial sector, such as the installation of solar rooftop, solar street light, and floating solar. Thus, industrial estates that gather a community of businesses located on a common property can realize more benefits of facilitation of renewable energy sources at the industrial level.

Like other countries, in Thailand, there are mainly three types of industrial production sites: (1) eco-industrial estate, (2) ordinary industrial estate, and (3) scattered home industry. Whereas scattered home industry means the individual manufacturers located in certain areas, ordinary industrial estate refers to an industrial site where numerous manufacturers share the same infrastructure and facilities. Eco-industrial estate is a community of industrial businesses located on a common property, and members seek enhanced environmental, economic, and social performance through collaboration in managing environmental and resource issues. Hence, eco-industrial estate aims restrict pollution and wastes within its boundary and emphasizes on environmental management and quality of life for the communities more than the others. On the other hand, scattered home industry cannot be easily controlled with respect to industrial pollution. Therefore, an eco-industrial estate is obviously more efficient with respect to the industrial pollution produced, but it requires high investment for environmental protection. An eco-industrial estate also targets pollution reduction and greenhouse gas reduction through low-carbon production.

This chapter looks at the implementation of renewable energy as a climate change mitigation measure in eco-industrial estates in Thailand in order to understand the existing situation as well as opportunities for renewable energy use. This chapter will illustrate eco-industrial estates in Thailand as a case study because the main achievement of eco-industrial estates in Thailand has been to make these industrial sites low-carbon sites, and Thai government has granted many concessions for greenhouse gas reduction initiatives.

1.2   THE PROMOTION OF RENEWABLE ENERGY IN THAILAND

Energy demand in Thailand is expected to increase by almost 80 percent in the next two decades, driven by massive amount of domestic and industrial consumption. In response, Thai government has sought to enhance its energy security by improving energy efficiency and promoting renewable energy. The Department of Alternative Energy Development and Efficiency (DEDE) under the Ministry of Energy has been established under the 1992 Energy Development and Promotion Act and has played a significant role in the promotion of renewable energy. Thailand has set a new target of 30 percent of renewable energy in the total final energy consumption by 2036 in its Alternative Energy Development Plan (AEDP) 2015. Renewable energy has high potential to be used in place of fossil energy, but the costs of renewable energy resources are still high compared with the costs of using commercial energy, particularly the development of solar energy, which requires the use of high-cost technology. However, solar energy is expected to be the main renewable energy source for electricity generation to the extent of 6,000 MW by the year 2036 (DEDE, 2015) because Thailand has a high potential for solar energy use in terms of concentration and amount of utilization areas due to its geographical location and landscape. Approximately 50 percent of Thailand's terrain is exposed to concentrated sunlight all year round (Sitdhiwej, 2005). As a result, to promote solar energy, the Board of Investment (BOI) provides financial subsidy in terms of tax, abatement, concessions, and partial subsidy for solar installations and the DEDE provides technology transfer and knowledge dissemination. In Thailand, solar farm, solar rooftop, solar street light, and floating solar are the preferred solar installations.

1.3 ECO-INDUSTRIAL ESTATES (EIE) IN THAILAND AND ITS CLIMATE CHANGE MITIGATION

Thailand had enthusiastically developed its industrial sector as it became the heart of Thai economy. At the same time, however, environmental quality and health of the communities had been ruined. To rectify the ignorance of environmental management on the part of industrial estates in causing environmental degradation and adverse impact on residential areas, the Thai government introduced the concept of eco-industrial estates (EIE) in 2000 (IEAT, 2012; Panyathanakun et al., 2013). The Industrial Estate Authority of Thailand (IEAT), as a state enterprise under the Ministry of Industry (MOI), was designated to develop and implement the appropriate scheme of EIE for emerging and existing industrial estates. The development of EIE in Thailand began with the initiation of "The Development of Eco Industrial Estates and Networks" or DEE + Net. Five industrial estates were selected as pilot locations for the introduction of various EIE concepts: the Northern Region, Map Ta Phut, Eastern Seaboard, Amata Nakorn, and Bang Pu industrial estates. These eco-industrial estates aimed to encourage industries to utilize and obtain value from waste through the concepts of 3Rs (reuse, reduce, and recycle). In 2004, the progress of the DEE +Net project was reported in the 2nd International Conference & Workshop for Eco Industrial Development held in Bangkok, Thailand. Among the project's reported successes were an increase in firms' awareness of EIE development concept, collaboration, and achieving financial benefits. However, barriers to further improvement were identified, such as tax issues relating to the movement of waste out of the free-trade zone, lack of government support and subsidy for clean technology, a lack of openness and efficient communication from IEAT, and shortcomings in the presentation of the EIE concept to firms (Lowe, 2001). Unfortunately, the first phase of EIE development ended in 2004 without further support. From the middle of 2006 until the present, EIE development, employing the eco-efficiency concept, was introduced to Map Ta Phut industrial estate. The development of appropriate eco-efficiency indicators for the evaluation of the industrial sector and industrial estates was a key objective of the EIE development project. The research contributed to the development of an EIE in Map Ta Phut industrial estate, the establishment of a researcher position at Map Ta Phut industrial estate for eco-efficiency indicator development, and collaboration between the industrial and academic sectors. In addition, it provided fundamental knowledge for eco-efficiency evaluation of the industrial sector in Thailand. Its framework was a useful tool for industrial assessment and would feed into strategic development and contribute to the establishment of additional EIE (Charmondusit and Keartpakpraek, 2011; Charmondusit, 2009).

Since 2010, due to severe environmental pollution and public opposition to industrial estates, IEAT has relaunched EIE development strategies with the novel introduction of explicit EIE indicators that included five aspects with 22 criteria to evaluate the performance of each EIE (Figure 1.1; IEAT, 2016a). The evaluation procedures have been set in more systematic patterns than in the previous phase. The IEAT has established the Department of Eco-industrial Town Development and trained EIE audit teams to respond in a specific matter in order to gain trust from the public. The initiatives developed are expected to be a crucial component of the further development of EIE and stakeholders' networks in Thailand. The IEAT had a plan to develop all the existing industrial estates in operation to EIE within the year 2021 (IEAT, 2016b).

An overview of each aspect is given below:

1. The physical aspect aims at achieving a proper landscaping plan and efficient infrastructure development under the EIE. The eco-design and green protection area are two of the focused areas.
2. The economic aspect aims at achieving growth and sustained positive economic results, as well as to encourage strengthening of the local economy, the wider community, and industry. Other target areas of this category include market development, transportation, and logistics.
3. The environmental aspect aims at encouraging the efficient use of resources, and effective emission and pollution mitigation. IEAT targets the management of waste water, solid waste, noise pollution, air pollution, efficient use of energy, eco-friendly processes, health and safety, eco-efficiency, and environmental monitoring.
4. The social aspect aims at encouraging a better quality of life for people who work in the estate and who live in surrounding communities.
5. The managerial aspect aims at establishing a systematic management process for the estate and facilitating continuous improvement. This will focus on collaboration among stakeholders, maintenance and improvement of the estate's management system, effective information and report management, and continuous improvement in the capability of personnel.

From the abovementioned criteria of eco-industrial estates, the implementation of eco-efficiency and renewable energy represents a part of climate change mitigation measures. Therefore it is necessary that all EIE have to response to climate change issue and establish mitigation measures to reduce greenhouse gas emissions.

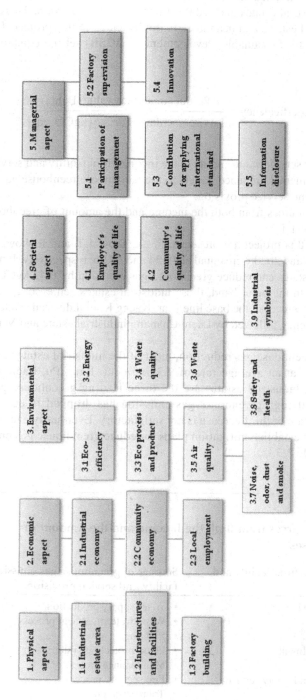

FIGURE 1.1 Eco-industrial estate characteristics in five aspects and with 22 criteria.

In 2018, IEAT initiated the project of eco-efficiency in order to assess eco-efficiency of industrial estates and reduce greenhouse gas emission. There were 13 pilot eco-industrial estates and headquarters implemented in this project. The World Business Council for Sustainable Development (WBCSB) set the equation of eco-efficiency as follows:

$$\text{Eco-effieciency} = \frac{\text{Product or service value}\left(\text{Baht}\right)}{\text{Environmental influence}\left(\text{kgCO}_2\text{eq}\right)}$$

where "product or service value" refers to the income from facility and service provision and "environmental influence" refers to the amount of greenhouse gas emission from the facility and service provision.

The details of sources from both the income and the amount of greenhouse gases are shown in Table 1.1.

The results of this project are presented in the Table 1.2, and it shows that eco-efficiency in 2018 and 2020 were equal at 48.54 and 48.96, respectively. It means that those industrial estates can reduce greenhouse gas emission by 0.42 kgCO$_2$eq per 1 baht of income. On the other hand, those industrial estates achieved a greater eco-efficiency of 0.86 percent in the base line year. Kaeng Khoi industrial estate achieved the best eco-efficiency followed by Laem Chabang industrial estate and Map Ta Phut industrial port.

In terms of greenhouse gas reduction, most of the industrial estates can reduce greenhouse gases at an average rate of 9.35 percent. Bang Pu, Laem Chabang, and Kaeng Khoi industrial estates are the three leading greenhouse gas reduction sites. Even though greenhouse gas emission has dropped in eco-industrial estates/port, the rate of this decrease is not satisfactory. Eventually each industrial estate/port would establish mitigation plans to reduce greenhouse gas emission as presented in Table 1.3.

TABLE 1.1
The Details of Sources from Both the Income and the Amount of Greenhouse Gases

Sources of income from facility and service provision	Sources of greenhouse gas emission from facility and service provision
• Maintenance service	• Water supply production
• Raw water	• Waste water treatment process
• Supply water	• Lighting
• Waste water treatment	• Flood protection system
• Waste collection	• Maintenance process
• Environmental and safety service	• Fire security process
• Operation service	• Transportation
	• Office operation

TABLE 1.2
The Result of the Project

No.	Industrial Estates/Port	Greenhouse gases emission (kgCO$_2$eq)			Income (Baht)			Eco-Efficiency (Baht/kgCO$_2$eq)		
		2018 (Based line)	2020	Difference	2018 (Based line)	2020	Difference	2018 (Based line)	2020	Difference
1	Kaeng Khoi	350,066	205,734	144,332	10,600,558	11,905,086	1,304,528	30.28	57.87	27.58
2	Pichit	182,520	241,577	−59,057	3,105,294	6,402,919	3,297,626	17.01	26.5	9.49
3	Southern	885,807	792,662	93,145	17,255,571	15,651,831	−1,603,740	19.48	19.75	0.27
4	Map Ta Phut	2,622,702	2,819,638	−196,936	1,335,465,527	1,220,717,851	−114,747,676	509.19	432.93	−76.26
5	Laem Chabang	4,499,132	3,204,593	1,294,540	283,554,153	267,076,087	−16,478,066	63.02	83.34	20.32
6	Northern Region	3,266,014	2,817,862	448,152	150,382,496	122,338,184	−28,044,312	46.04	43.42	−2.63
7	Lad Krabang	17,158,067	16,244,145	913,922	291,079,336	268,900,662	−22,178,674	16.96	16.55	−0.41
8	Bangchan	1,838,031	1,924,366	−86,334	38,159,320	38,409,525	250,204	20.76	19.96	−0.8
9	Samut Sakhon	6,466,793	6,332,339	134,455	200,230,726	170,359,054	−29,871,672	30.96	26.9	−4.06
10	Bang Pu	16,054,150	13,544,929	2,509,220	285,819,681	258,932,498	−26,887,183	17.8	19.12	1.31
11	Bang Phli	964,782	953,717	11,064	28,229,804	26,832,104	−1,397,700	29.26	28.13	−1.13
12	Map Ta Phut Industrial Port	834,829	672,775	162,054	39,772,445	38,731,741	−1,040,704	47.64	57.57	9.93
13	Nakhon Luang	752,952	895,547	142,596	28,614,061	33,451,316	4,837,255	38	37.35	−0.65
14	Headquarters	1,399,634	1,069,162	330,472						
	Total	55,875,845	50,649,884	5,225,961 / 9.35%	2,712,268,971	2,479,708,858	−232,560,113 / −8.57%	48.54 / Factor	48.96 / 1.009	0.42 / 0.86%

Note: 1 US dollar = 30.36 Thai baht (updated on March 3, 2021).

TABLE 1.3
Mitigation Plan to Reduce Greenhouse Gases

Mitigation plans	The amount of industrial estates/ port participation	The amount of greenhouse gases reduction (kgCO$_2$eq)
1. Replacing energy saving equipment		
1.1 Using LED lighting	13	309,326
1.2 Changing air conditioner	4	14,344
1.3 Changing water pump	7	41,657
2. Installing energy saving equipment		
2.1 Installing inverter pump	4	187,106
2.2 Installing other saving equipment	2	15,388
3. Installing solar energy	6	219,785
4. Controlling employee's behavior by setting schedule of electricity switch on and off and practicing green meeting	10	163,398
5. Reducing waste and water consuming		
5.1 Saving water	4	24,388
5.2 3Rs	2	234,684
5.3 Changing type of chemical used in water supply production	2	29,620
Total	54	1,239,696

1.4 FURTHERING RENEWABLE ENERGY IN ECO-INDUSTRIAL ESTATES

Adapting and utilizing renewable energy become the crucial indicator of eco-industrial estate performance in Thailand so that those eco-industrial estates will be lauded for their dedication to low-carbon initiatives. In the past, utilizing renewable energy has been promoted for consuming electricity supply from renewable energy generators using bio-diesel and bio-gasoline. However, in the recent years, many eco-industrial estates have shifted to self-producing renewable energy from sunlight. Solar street light, solar rooftop, and floating solar have been adapted for popular use in their utilities. In Thailand, solar energy has the potential for significant growth because it is a key renewable green energy for development, supported by Thai government policies. Solar energy is a key feature in Thailand's Alternative Energy Development Plan 2015–2036, which proposes a target of 30 percent renewable energy in the total energy consumption by 2036. The solar energy market also has been opening up for private investment, operations, and innovation, and is part of the recent solar regulations. This will promote private electricity generation, especially in industrial and commercial buildings. Solar radiation becomes a key factor of solar energy. DEDE (2019) reported that yearly average of daily global solar radiation during 2018–2020 is equal to 17.6–17.8 MJ/m^3-day, which has a high efficiency to be converted into energy.

For eco-industrial estates, in the past, the installation of solar energy for generating electricity was barely applied because of high investment and the instability of energy. However in recent years, solar energy has achieved significant growth due to the development of technology, low cost, and favorable regulations. Solar energy produced in industrial estates mostly generate electricity for their self-consumption and not for sale. Additionally, it has been proved that solar reduces greenhouse gas emission as well as lowers electricity cost. The preferred solar energy application in eco-industrial estates would be solar street light and solar rooftop because they are suitable for installation within the facilities and accommodations in eco-industrial estates.

In recent years, a solar energy innovation named "floating solar" as shown in Figure 1.2 is increasingly adapted for installation in eco-industrial estates. Whereas almost all the eco-industrial estates have to provide the retention ponds for holding discharged water and their own reservoir to keep raw water, floating solar has become an appropriate source of renewable technology. Installing floating solar extends the utmost benefit of using the area in retention ponds or reservoirs. According to the technical features of solar panels, installing solar panels on water surface increases the efficiency of solar panels due to the lower surface temperature. By installing solar panels on water, which has a lower temperature than on land, it can increase the efficiency of power generation from solar panels by around 10 percent. It can be used to produce power for self-consumption and also reduces greenhouse gas emissions. The benefits of floating solar can also help reduce evaporation, and it doesn't take up valuable space on land, which can be used for other purposes. The advantages of floating solar have been obviously demonstrated, so the Thai government intends to promote it by offering subsidy or concessions for installing floating solar. Floating solar capacity is rapidly growing from 70 megawatts of peak power (MWp) in 2015

FIGURE 1.2 Floating solar. *Source*: SCG (2021).

to 1,300 MWp in 2018. Currently, there are more than 300 floating solar installations worldwide and it is estimated that global demand for floating solar power is expected to grow by 22 percent year-on-year on average from 2019 through 2024 (IFC, 2020). Therefore, the facility of self-generating renewable energy by installing floating solar increasingly spread throughout eco-industrial estates. Within eco-industrial estates, a target has been set to reduce greenhouse gas emission by 2,500,000 $kgCO_2eq$ by the year 2025 (IEAT, 2020).

1.5   CONCLUDING REMARKS

Climate change has made the world awaken to its impacts. The United Nations stressed this issue in the 13th goal of SDGs and requires the nations to act toward solutions. Industrial estates are the powerful source of greenhouse gas emissions because industrial production sites are concentrated in a certain location. Therefore, the concept of eco-industrial estate has been implemented for achieving low-carbon emission as well as for forming a closed loop society. Promoting and applying eco-efficiency and renewable energy are the mitigation strategies to reduce greenhouse gases for eco-industrial estates in Thailand. The IEAT has launched the project of "eco-efficiency" to control, manage, and evaluate greenhouse gas emission within industrial estates and projected to reduce 2,500,000 $kgCO_2eq$ of greenhouse gases by 2025. The shift of consuming renewable energy to self-producing renewable energy provides a massive change for greenhouse gas reduction within eco-industrial estates. Thus, renewable energy from sunlight is increasingly as the source for self-producing renewable energy; however, floating solar is the suitable technology for eco-industrial estates because of the presence of retention ponds and reservoirs in most of those estates. The following significant benefits of floating solar also have been observed:

- It does not need valuable space on land so that the landlord can utilize the property for other purposes.
- Water bodies exert a cooling effect, which improves the performance efficiency of solar panels by 5–10 percent.
- It reduces shading, decreases civil works, lessens grid interconnection costs, minimizes water evaporation, and shortens algal blooming; in these ways, it improves water quality.
- As the demand for floating solar increases, the cost is likely to come down.

Solar radiation in Thailand is also conducive for installing floating solar; therefore, generating renewable energy through floating solar in eco-industrial estates is likely grow as an important part of the effort to address climate change in Thailand.

REFERENCES

Charmondusit, K. (2009). Eco-efficiency analysis and development of enterprise in Rayong province. *Area Based Development Research Journal* 2 (2), 5–16 (in Thai).
Charmondusit, K., and Keartpakpraek, K. (2011). Eco-efficiency evaluation of the petroleum and petrochemical group in the Map Ta Phut Industrial Estate, Thailand. *Journal of Cleaner Production* 19, 241–252.

DEDE. (2015). Alternative Energy Development Plan: AEDP2015. Department of Renewable Energy Development and Energy Efficiency, Ministry of Energy, Thailand.

DEDE. (2019). Thailand Alternative Energy Situation. Department of Alternative Energy Development and Efficiency, Ministry of Energy, Thailand.

IEAT. (2012). Standard and Guideline for Eco-industrial Estate. Industrial Estate Authority of Thailand (in Thai).

IEAT. (2016a). Characteristics, Indicators and Targets of Eco-industrial Estate in Thailand. Industrial Estate Authority of Thailand (in Thai).

IEAT. (2016b). The 5-Year IEAT Organizational Strategies (2017–2021). Industrial Estate Authority of Thailand (in Thai).

IEAT. (2020). The Evaluation of Eco-Efficiency Report. Industrial Estate Authority of Thailand (in Thai).

IFC. (2020). *Floating Solar Photovoltaic on the Rise*. World Bank Group, Washington, DC.

IPCC. (2014). Summary for policymakers. In: *Climate Change 2014: Mitigation of Climate Change. Contribution of Working Group III to the Fifth Assessment Report of the Intergovernmental Panel on Climate Change.* O. Edenhofer, R. Pichs-Madruga, Y. Sokona, E. Farahani, S. Kadner, K. Seyboth, A. Adler, I. Baum, S. Brunner, P. Eickemeier, B. Kriemann, J. Savolainen, S. Schlömer, C. von Stechow, T. Zwickel, and J.C. Minx (eds.). Cambridge University Press, Cambridge, 1–32.

IPCC. (2018a). Summary for policymakers. In: *Global Warming of 1.5°C. An IPCC Special Report on the Impacts of Global Warming of 1.5°C above Pre-industrial Levels and Related Global Greenhouse Gas Emission Pathways, in the Context of Strengthening the Global Response to the Threat of Climate Change, Sustainable Development, and Efforts to Eradicate Poverty.* V. Masson-Delmotte, P. Zhai, H.-O. Pörtner, D. Roberts, J. Skea, P.R. Shukla, A. Pirani, W. Moufouma-Okia, C. Péan, R. Pidcock, S. Connors, J.B.R. Matthews, Y. Chen, X. Zhou, M.I. Gomis, E. Lonnoy, T. Maycock, M. Tignor, and T. Waterfield (eds.). World Meteorological Organization, Geneva, Switzerland, 1–24.

IPCC (2018b). *Global warming of 1.5°C: an IPCC Special Report on the impacts of global warming of 1.5°C above pre-industrial levels and related global greenhouse gas emission pathways.* In: *The Context of Strengthening the Global Response to the Threat of Climate Change, Sustainable Development, and Efforts to Eradicate Poverty.* V. Masson-Delmotte, P. Zhai, H.O. P€ortner, D. Roberts, J. Skea, P.R. Shukla, A. Pirani, W. Moufouma-Okia, C. P_ean, R. Pidcock, S. Connors, J.B.R. Matthews, Y. Chen, X. Zhou, M.I. Gomis, E. Lonnoy, T. Maycock, M. Tignor, T. Waterfield (eds.). World Meteorological Organization, Geneva, Switzerland, 320–326.

Lowe, E. A. (2001). Eco-industrial parks handbook for Asian developing country. In: *2nd International Conference & Workshop for Eco-Industrial Development*, March 7–12, 2004, Thailand.

Nyong, A., Adesina, F., and Osman Elasha, F. (2007). The value of indigenous knowledge in climate change mitigation and adaptation strategies in the African Sahel. *Journal of Mitigation Adaptation Strategy Global Change* 12, 787–797.

Panyathanakun, V., Tantayanon, S., Tingsabhat, C., and Charmondusit, K. (2013). Development of eco-industrial estates in Thailand: Initiatives in the northern region community-based eco-industrial estate. *Journal of Cleaner Production* 51, 71–79.

SCG. (2021). *SCG Floating Solar Solutions*. From website: www.scg.com/innovation/en/scg-floating-solar. Siam Cement Group, Thailand.

Sitdhiwej, C. (2005). Paper presented at the TC Beirne School of Law Postgraduate Law Research Colloquium, Brisbane, December 2–4, 2005.

UNESCO. (2019). *Climate Change Mitigation and Adaptation: Simple Guide to School in Africa*. The United Nations Educational, Scientific and Culture Organization, Nairobi, Kenya.

Vijaya Venkataraman, S., Iniyan, S., and Goic, R. (2012). A review of climate change, mitigation and adaptation. *Renewable and Sustainable Energy Reviews* 16, 878–897.

Xinsheng, L., Arnold, V., and Letitia, A. (2008). Regional news portrayals of global warming climate change. *Environmental Science & Policy* 11(5), 379–393.

2 Estimation of Global Solar Energy to Mitigate World Energy and Environmental Vulnerability

Md. Faruque Hossain

HIGHLIGHTS

- Estimation of net fossil fuel reserves on Earth
- Calculation of the annual global energy demand
- Calculation of the annual global solar energy reaching Earth
- Estimation of the net electricity power generation from the net solar energy reaching Earth annually
- Solar energy implementation to satisfy the world's energy requirements.

2.1   INTRODUCTION

The conventional global energy consumption for powering modern civilization is indeed accelerating the finite level of the current fossil fuel reserves of 36,630 EJ (12,24). The global fossil fuel energy consumption was 283 EJ/Yr in 1980, 348 EJ/Yr in 1995, 405 EJ/Yr in 2005, and 515 EJ/Yr in 2015 and will reach 610 EJ/Yr in 2025, 705 EJ/Yr in 2035, 860 EJ/Yr in 2045, and 990 EJ/Yr in 2050 (1–3). The

DOI: 10.1201/9781003240129-2

global utilization of fossil fuels in 2018 was 2.236×10^{20} EJ, which was responsible for releasing 8.01×10^{11} tons of CO_2 into the atmosphere and rapidly accounts for the acceleration of deadly climate change (4–6). Consequently, adverse environmental impacts, such as acidic rain, floods, and climate change, are occurring unpredictably throughout the world (7–9). A recent study showed that the current concentration of CO_2 in the atmosphere is 400 ppm, which needs to be lowered to a standard level grade of 300 ppm CO_2 for the wellness of clean breathing and respiratory systems for all mammals (10–12). Another study revealed that greenhouse gas emissions rapidly accelerate fluctuation of the global diurnal mean temperature, posing a serious threat to the natural ecosystems and human well-being due to the utilization of fossil fuels, since they create radioactive CO_2 into the atmosphere over a certain period of time (13,14).

Unfortunately, the consumption of conventional energy is still accelerating rapidly throughout the world, and the situation will remain unchanged until a renewable source of energy is developed to utilize sustainable energy (15–17). Simply put, there is an urgent demand to develop sustainable energy technology to mitigate fossil fuel consumption where the "new source … fulfills the requirement of the present without disrupting the ability of future demand of energy to fulfill the complete needs for future generations." Hence, global solar energy utilization is an interesting source to fulfill the net energy requirement throughout the world. It is a natural renewable energy source that is generated by nuclear fusion in the sun, and is constantly flowing away from the sun toward the solar system, with part of it reaching Earth. It is a tremendous source of clean and renewable energy (18–20). If only 0.001 percent of the annual solar energy reaching Earth was used, which is clean and abundant everywhere, it would meet the net energy demand for the entire planet. Therefore, in this study, research was performed to harvest the total global solar energy reaching Earth, which is clean and environmentally friendly, in order to mitigate the global net energy needs.

2.2 MATERIALS, METHODS, AND SIMULATION

2.2.1 CALCULATION OF THE NET SOLAR ENERGY ON EARTH

Earth's total surface is clarified by characterizing various directional angles considering the Cartesian coordinate system, where x denotes the skyline convention, y denotes east–west, and z denotes the zenith, in order to measure the total solar irradiance during any day of the year (Figure 2.1). The position of the celestial body in this framework is thus chosen by h, which is denoted for height, and A denotes the azimuth angle, while the central framework utilizes it as the convention factor, which is the z hub. It focuses on the North Pole, the y hub indistinguishably focuses on the horizon of the skylight, and x pivots opposite to both the North Pole and the horizon. Therefore, the angles and coordinate frequencies are encountered mathematically by calculating the latitude and longitude in order to implement correct angles to trap the solar irradiance most efficiently. Here, the zero point of the latitude is considered the primary meridian, which controls the function of the meridian of the Eastern Hemisphere and the Western Hemisphere angle of the Earth's surface. North

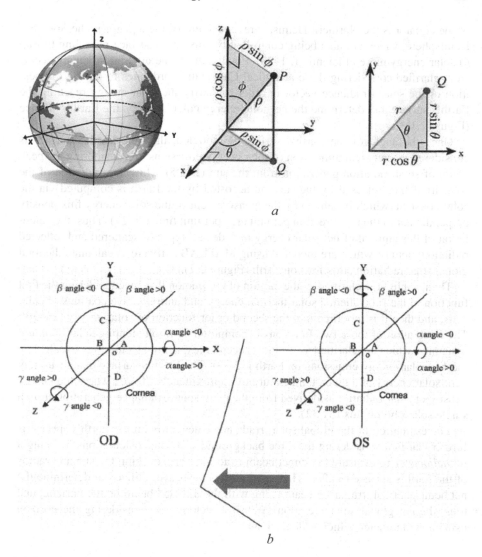

FIGURE 2.1 (a) Cartesian coordinates clarification of the southern axis *x*, western axis *y*, and the zenith axis *z* to calculate the total solar energy reaching Earth, considering that the average energy density of sunlight on the surface of Earth is 1,366 W/m² by implementing the diameter of Earth as 10,000,000 of the meridian at the North Pole to the equator and the radius of Earth is $2/\pi \times 10^7$ m. The location of this celestial body is analyzed by determining two angles, sin θ and cos θ. (b) The longitudinal and latitudinal equatorial angles have been clarified, where the convention *z*-axis point denotes the North Pole, and the east–west axis *y*-axis denotes the identical angle of the horizon.

of the equator is the Northern Hemisphere, and south of the equator is the Southern Hemisphere, which are also being controlled by this Earth surface modeling to trap to solar energy more efficiently (21–23). Finally, the δ and ω point hours are accurately clarified considering these analytical Cartesian coordinates to decide the position of the solar irradiance vector in order to clarify the solar energy reaching the Earth's surface and determine the net solar energy calculation on the Earth's surface (Figure 2.1).

Once the angle of the Earth's surface is modeled, then the Earth's surface is considered the net areal dimension of solar energy emissions by considering the peak hours of solar radiation generation from the sun (24,25). Thus, the radiation of the solar irradiance released by the sun and accosted by the Earth is computed via the solar constant, which is defined by the measurement of the solar energy flux density perpendicular to the ray direction per unit area per unit time (26,27). Thus, the calculation of this amount of net solar energy includes all types of scattered and reflected radiation, both of which are modeled using MATLAB software to calculate the total global solar radiation emissions on Earth (Figure 2.2).

Then, sunlight is clarified as the motion of the photon flux by considering the first function of the fundamental solar thermal energy and antireflective coatings of solar cells, and then it is modified into the second order function of solar energy (28–30). The integration of these two functions is computed by implementing solar quantum dynamics, which are clarified as the most acceptable quantum technology to calculate the net solar energy emissions on Earth (12,33). The Earth's surface can accurately emit solar energy at a given temperature of approximately 700°C, where the energy density of solar radiation is derived from the peak solar irradiance generation from a single solar photon flux (31,32).

The estimation of the global solar irradiance calculation on the Earth's surface is further clarified considering the three background solar data calculations by using a *pyrheliometer* to measure the direct beam irradiance approaching the sun and radius of the Earth's surface (34,35). Then, a *pyranometer* is also utilized to determine the net hemispherical irradiance beam along with the diffused beam on the horizon, and thus, the net global solar radiation (W/m²) is determined considering the horizon using a pyranometer, which is denoted as:

$$I_{tot} = I_{beam} \cos\theta + I_{diffuse} \qquad (2.1)$$

where θ represents the zenith angle, which has been implemented to calculate the net solar energy reaching Earth (Figure 2.2).

This measurement is then confirmed against modest pyrheliometers using the thermocoupler detector and the PV detector recorder considering the determination of the wavelength of the solar spectrum (36,37). Eventually, a photoelectric sunshine recorder is used, which is intermittent and varies by the solar irradiance intensity (32,38). Since solar radiation is related to the photon charge, the attributes of the photon energy on the Earth's surface are computed considering the quantum flow of photon radiation on a global scale by using classical multidimensional scaling in MATLAB 9.0 (31,32). Consequently, a computational model of the photon radiation

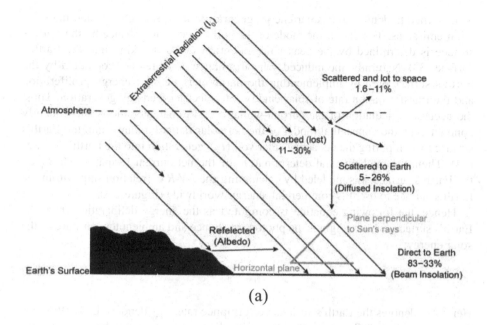

(a)

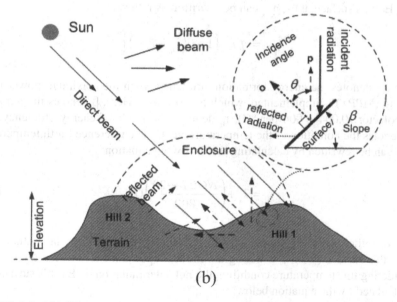

(b)

FIGURE 2.2 (a) The emission of solar energy on the Earth's surface, (b) various types of irradiance on the Earth's surface; direct beam, reflected beam, diffuse beam at various angles.

is quantified to demonstrate solar energy generation from sunlight considering radiation emissions. Thereafter, the mode of the solar quanta absorbance by the Earth's surface is determined by the peak solar radiation output tracking into the Earth's surface (33). Naturally, the induced solar irradiance is, thereafter, computed by the Earth's surface area by implementing the parameters of solar energy proliferation and the transformation rate of solar energy into electricity energy generation. Thus, the accurate calculation of the current–voltage (I–V) characteristic is subsequently conducted by the conceptual model of the net solar radiation intake into the Earth's surface by computing the net active solar volt (I_{v+}) generation into the Earth's surface (34). Then, the mathematical determination of the net current formation via I_{pv} on the Earth's surface was modeled by calculating the I-V-R correlation ship within the Earth's surface to use this commercial energy worldwide (Figure 2.3).

Hence, the following equation is computed as the energy deliberation from the Earth's surface, whose origin is the photon irradiance and ambient temperature of the solar energy:

$$P_{pv} = \eta_{pvg} A_{pvg} G_t \qquad (2.2)$$

Here, η_{pvg} denotes the Earth's surface performance rate, A_{pvg} denotes the Earth's surface array (m²), and G_t denotes the photon irradiance intake rate on the plane (W/m²) of the Earth's surface; thus, η_{pvg} can be rewritten as follows:

$$\eta_{pvg} = \eta_r \eta_{pc} \left[1 - \beta\left(T_c - T_{c\,ref}\right)\right] \qquad (2.3)$$

where η_{pc} denotes the energy formation efficiency when the maximum power point tracking (MPPT) is implemented, which is close to 1. Here, β denotes the temperature cofactor (0.004–0.006 per °C), η_r denotes the mode of energy efficiency, and T_{cref} denotes the condition of the temperature at °C. The reference Earth temperature ($T_{c\,ref}$) can be rewritten by calculating the following equation:

$$T_c = T_a + \left(\frac{NOCT - 20}{800}\right) G_t \qquad (2.4)$$

T_a denotes the encircling temperature in °C, and G_t denotes the solar radiation on the Earth's surface (W/m²), denoting the modest optimum Earth temperature in °C. Considering this temperature condition, the net solar radiation on Earth's surface can be calculated by the equation below:

$$I_t = I_b R_b + I_d R_d + \left(I_b + I_d\right) R_r \qquad (2.5)$$

Solar energy necessarily works as a conceptual P–N junction superconductor to form electricity though the Earth's surface, which is interlinked in a parallel series connection (45). Thus, a unique conceptual circuit model with respect to

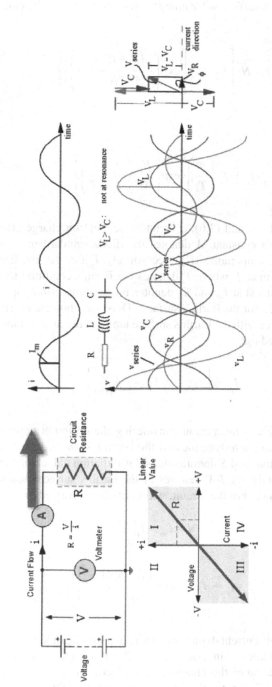

FIGURE 2.3   The conceptual circuit diagram of the whole Earth's surface depicts the net photophysical current generation into the Earth's surface by detailing the model of the *I-V-R* relationship to use electricity throughout the world.

the N_s series of Earth's surface and the N_p parallel arrays has been computed by the following the Earth surface solar energy equation based on the current and volt relationship:

$$I = N_p \left[I_{ph} - I_{rs} \left[\exp\left(\frac{q(V + IR_s)}{AKTN_s} \right) - 1 \right] \right] \tag{2.6}$$

where

$$I_{rs} = I_{rr} \left(\frac{T}{T_r} \right)^3 \exp\left[\frac{E_G}{AK} \left(\frac{1}{T_r} - \frac{1}{T} \right) \right] \tag{2.7}$$

Hence, in Equations (2.5) and (2.6), q denotes the electron charge (1.6×10^{-19} C), K denotes Boltzmann's constant, A denotes the diode standardized efficiency, and T denotes the Earth's temperature (K). Accordingly, I_{rs} denotes the Earth's surface reversed current motion at T, where T_r denotes the Earth's condition temperature, I_{rr} denotes the reverse current at T_r, and E_G denotes the photonic band-gap energy of the superconductor utilized for the Earth's surface. Thus, the photonic current I_{ph} will be generated in accordance with the Earth's surface temperature and radiation condition, which can be expressed by:

$$I_{ph} = \left[I_{SCR} + k_i \left(T - T_r \right) \frac{S}{100} \right] \tag{2.8}$$

Here, I_{SCR} denotes the current motion considering the optimum temperature of the Earth and the solar radiation dynamics on the Earth's surface, k_i denotes the short-circuited current motion, and S denotes the solar radiation calculation in a unit area (mW/cm²). Subsequently, the I–V features of the Earth's surface shall be deformed from the conceptual model of the circuit, which can be expressed as:

$$I = I_{ph} - I_D \tag{2.9}$$

$$I = I_{ph} - I_0 \left[\exp\left(\frac{q(V + R_s I)}{AKT} \right) - 1 \right] \tag{2.10}$$

I_{ph} denotes the photonic current dynamics (A), I_D denotes the diode-originated current dynamics (A), I_0 denotes the inversed current dynamics (A), A denotes the diode-induced constant, q denotes the charge of the electron (1.6×10^{-19} C), K denotes Boltzmann's constant, T denotes the Earth's temperature (°C), R_s denotes the series resistance (ohm), R_{sh} denotes the shunt resistance (Ohm), I denotes the cell current motion (A), and V denotes the Earth voltage motion (V). Therefore, the net current flow into the Earth's surface can be determined using the following equation:

$$I = I_{PV} - I_{D1} - \left(\frac{V + IR_S}{R_{SH}}\right) \tag{2.11}$$

where

$$I_{D1} = I_{01}\left[\exp\left(\frac{V + IR_s}{a_1 V_{T1}}\right) - 1\right] \tag{2.12}$$

Here, I and I_{01} denote the reversed current flow into the conceptual circuit, and V_{T1} and V_{T2} denote the optimum thermal volts into the circuit. Thus, the circuit standard factors are presented by a_1 and a_2, and then the mode of the Earth's surface is normalized by expressing the following equation:

$$v_{oc} = \frac{V_{oc}}{cKT/q} \tag{2.13}$$

$$P_{max} = \frac{\dfrac{V_{oc}}{cKT/q} - \ln\left(\dfrac{V_{oc}}{cKT/q} + 0.72\right)}{\left(1 + \dfrac{V_{oc}}{KT/q}\right)}\left(1 - \dfrac{V_{oc}}{V_{oc}}\right)\left(\dfrac{V_{oc0}}{I_{SC}}\right)\left(\dfrac{V_{oc0}}{1 + \beta\ln\dfrac{G_0}{G}}\right)\left(\dfrac{T_0}{T}\right)^y I_{sc0}\left(\dfrac{G}{G_0}\right)^a \tag{2.14}$$

where v_{oc} denotes the standard point of the open-circuit voltage V_{oc}, the thermal voltage $V_t = nkT/q$, c denotes the constant current motion, K denotes Boltzmann's constant, T denotes the temperature into the Earth's surface PV cell in Kelvin, α denotes the function that represents the nonlinear motion of photocurrents, q denotes the electron charge, y denotes the function acting for all nonlinear temperature–voltage currents, and β denotes the Earth surface mode for a specific dimensionless function to enhance the current flow rate. Subsequently, Equation (2.14) represents the peak energy generation from the Earth surface module, which is interlined in both series and parallel connections. Thus, the equation for the net energy formation into the array of N_s has been interlinked in series, and N_p has been interlinked in parallel considering the power P_M of each mode of connection, which is finally expressed using the following equation:

$$P_{array} = N_s N_p P_M \tag{2.15}$$

2.2.2 MODELING OF NET ELECTRICITY ENERGY GENERATION FROM THE TOTAL SOLAR IRRADIANCE ON EARTH

To convert the global solar energy into electricity energy, a model is also being prepared by integrating global Albanian symmetries of the gauge field scalar (32,36). Naturally, the net solar energy particle will functionally act as the dynamic photons of particle T^α at the global symmetrical array of Earth's surface by initiating the gauge field of $A_\mu^\alpha(x)$, and then the local Albanian will subsequently be started to activate at the global $U(1)$ phase symmetry to deliver the net electricity energy (21,37). Thus, the model is considered a complex vector field of $\Phi(x)$ of the Earth's surface, where the electric charge q will couple with the EM field of $A^\mu(x)$; thus, the equation can be expressed by $\mathfrak{h}$:

$$\mathfrak{h} = -\frac{1}{4}F_{\mu\nu}F^{\mu\nu} + D_\mu\Phi^* D^\mu\Phi - V(\Phi^*\Phi) \qquad (2.16)$$

where

$$D_\mu\Phi(x) = \partial_\mu\Phi(x) + iqA_{\frac{1}{4}}(x)\Phi(x)$$

$$D_\mu\Phi^*(x) = \partial_\mu\Phi^*(x) - iqA_{\frac{1}{4}}(x)\Phi^*(x) \qquad (2.17)$$

and

$$V(\Phi^*\Phi) = \frac{\lambda}{2}(\Phi^*\Phi)^2 + m^2(\Phi^*\Phi) \qquad (2.18)$$

Here, $\lambda > 0$ $m^2 < 0$; therefore, $\Phi = 0$ is a local optimum vector quantity, while the minimum form of the degenerated scalar circle is clarified as $\Phi = \frac{v}{\sqrt{2}} * e^{i\theta}$,

$$v = \sqrt{\frac{-2m^2}{\lambda}}, \text{ any real } \theta \qquad (2.19)$$

Subsequently, the vector field Φ of the Earth's global surface will form a nonzero functional value $\Phi \neq 0$, which will simultaneously determine the $U(1)$ symmetrical net solar energy generation. Therefore, the global $U(1)$ net symmetrical electrical energy of $\Phi(x)$ will be delivered as the expected value of Φ by confirming the x-dependent state of the symmetrical $\Phi(x)$ array of the Earth's surface and can be expressed by the following equation:

$$\Phi(x) = \frac{1}{\sqrt{2}}\Phi_r(x) * e^{i\Theta(x)}, \text{real } \Phi_r(x) > 0, \text{real } \Phi(x) \qquad (2.20)$$

Thus, the net calculation of the electricity energy generation from the net Earth's surface solar energy is being determined by considering the vector $\Phi\left(x\right)=0$, and it is a first-order function of $\Phi \neq 0$, considering the peak level of solar energy emission on the Earth's surface of $\Phi x \neq 0$. Thus, the net electricity energy generation from the global solar energy calculation $\phi_r\left(x\right)$ and $\Theta(x)$ and its vector on the Earth surface field ϕ_r has been confirmed by the following equation:

$$V\left(\phi\right)=\frac{\lambda}{8}\left(\phi_r^2 - v^2\right)^2 + \text{const,} \tag{2.21}$$

or the resultant electricity energy generation is shifted by its VEV, $\Phi_r\left(x\right)=v+\sigma\left(x\right)$:

$$\phi_r^2 - v^2 =\left(v+\sigma\right)^2 - v^2 = 2v\sigma+\sigma^2 \tag{2.22}$$

$$V=\frac{\lambda}{8}\left(2v\sigma-\sigma^2\right)^2 =\frac{\lambda v^2}{2}*\sigma^2 +\frac{\lambda v}{2}*\sigma^3 +\frac{\lambda}{8}*\sigma^4 \tag{2.23}$$

Simultaneously, the functional derivative $D_\mu\phi$ will become:

$$D_\mu\phi=\frac{1}{\sqrt{2}}\left(\partial_\mu\left(\phi_r e^{i\Theta}\right)+iqA_\mu * \phi_r e^{i\Theta}\right)=\frac{e^{i\Theta}}{\sqrt{2}}\left(\partial_\mu\phi_r +\phi_r * i\partial_\mu\Theta+ \phi_r *iqA_\mu\right) \tag{2.24}$$

$$\left|D_\mu\phi\right|^2 =\frac{1}{2}\left|\partial_\mu\phi_r +\phi_r * i\partial_\mu\Theta+ \phi_r *iqA_\mu\right|^2$$

$$=\frac{1}{2}\left(\partial_\mu\phi_r\right)+\frac{\phi_r^2}{2}*\left(\partial_\mu\Theta qA_\mu\right)^2$$

$$=\frac{1}{2}\left(\partial_\mu\sigma\right)^2 +\frac{\left(v+\sigma\right)^2}{2}*\left(\partial_\mu\Theta+qA_\mu\right)^2 \tag{2.25}$$

Altogether,

$$\mathcal{h}=\frac{1}{2}\left(\partial_\mu\sigma\right)^2 -v\left(\sigma\right)-\frac{1}{4}F_{\mu v}F^{\mu v} +\frac{\left(v+\sigma\right)^2}{2}*\left(\partial_\mu\Theta+qA_\mu\right)^2 \tag{2.26}$$

To determine the formation of this net electricity generation referred to as $(\mathcal{h}_{sef})$ into the Earth's surface, the function of the electrostatic fields has been quantified by conducting the quadratic calculation and is described by the following equation:

$$\mathfrak{H}_{sef} = \frac{1}{2}\left(\partial_\mu \sigma\right)^2 - \frac{\lambda v^2}{2} * \sigma^2 - \frac{1}{4}F_{\mu\nu}F^{\mu\nu} + \frac{v^2}{2} * \left(qA_\mu + \partial_\mu\Theta\right)^2 \qquad (2.27)$$

Here, this net electricity generation ($\mathfrak{H}_{free}$) function will certainly admit a realistic vector particle of positive mass^2 = λv^2 integrating the areal $A_\mu(x)$ function and the electricity energy generation fields $\Theta(x)$ to confirm the calculation of the net electricity energy from the global solar energy onto the Earth's surface (12,28).

2.3   RESULTS AND DISCUSSION

2.3.1   CALCULATION OF THE NET SOLAR ENERGY ON EARTH

To calculate the net solar energy on the Earth's surface, the net irradiance of photon emissions was calculated by integrating Equations (2.25) and (2.26). Necessarily, the functional Earth surface area $J(\omega)$, the photonic quantum field and the unit area $J(\omega)$ are calculated by considering the constant irradiance coupling point and the Weisskopf–Winger approximation mechanism to confirm the accurate solar energy reaching the Earth's surface (Figure 2.4).

The computed results show that the distribution of solar irradiance on the Earth's sphericity and orbital parameters is the application of the unidirectional beam incident to a rotating sphere of Milankovitch cycles from the spherical Earth law of cosines:

$$\cos(c) = \cos(a)\cos(b) + \sin(a)\sin(b)\cos(C) \qquad (2.28)$$

where a, b, and c are considered the arc lengths, in radians, of the sides of a spherical triangle, and C represents the angle of the vertex that has an arc length of c. To determine the solar zenith angle Θ, the following equation is clarified considering the application of the spherical law of cosines:

$$C = h$$

$$c = \Theta$$

$$a = \frac{1}{2}\pi - \phi$$

$$b = \frac{1}{2}\pi - \delta$$

$$\cos(\Theta) = \sin(\phi)\sin(\delta) + \cos(\phi)\cos(\delta)\cos(h) \qquad (2.29)$$

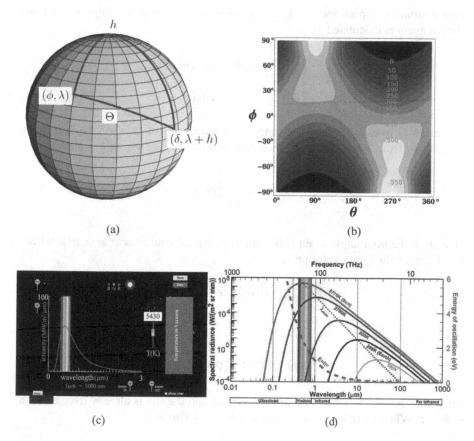

FIGURE 2.4 (a) The sphere triangular for cosines is clarified as the solar zenith angle Θ considering latitude φ and longitude λ; (b) the average daily irradiation at the top of the atmosphere, where θ is the polar angle of the Earth's orbit, θ = 0 at the vernal equinox, and θ = 90° at the summer solstice, while φ is the latitude of the Earth; (c) shows the solar irradiance at various frequencies, and (d) is the peak temperature, which suggests the calculative power to determine the net solar energy.

To simplify this equation, it has been further clarified as a general derived as follows:

$$
\begin{aligned}
\cos(\theta) = {} & \sin(\phi)\sin(\delta)\cos(\beta) + \sin(\delta)\cos(\phi)\sin(\beta)\cos(\gamma) \\
& + \cos(\phi)\cos(\delta)\cos(\beta)\cos(h) - \cos(\delta)\sin(\phi)\sin(\beta)\cos(\gamma)\cos(h) \\
& - \cos(\delta)\sin(\beta)\sin(\gamma)\sin(h)
\end{aligned} \tag{2.30}
$$

where β denotes the angle from the horizon and γ denotes the azimuth angle.

The sphere of Earth from the sun here is denoted by R_{E}, where the average distance is represented as R_0, with an approximation of one astronomical unit (AU). Here, the solar constant is represented as S_0, where the solar irradiance density onto an Earth plane tangent is calculated as:

$$Q = \begin{cases} S_o \dfrac{R_0^2}{R_E^2}\cos(\theta) & \cos(\theta) > 0 \\[2ex] 0 & \cos(\theta) \leq 0 \end{cases} \tag{2.31}$$

The mean Q over a day is the average of Q over one rotation, or the hour angle progressing from $h = \pi$ to $h = -\pi$. Thus, the equation has been rewritten as:

$$Q^{-\text{day}} = -\frac{1}{2\pi}\int_{\pi}^{-\pi} Q dh \tag{2.32}$$

Since h_0 is the hour angle when Q becomes positive, it could occur at sunrise when $\Theta = 1/2\,\pi$, or for h_0 as a solution of:

$$\sin(\phi)\sin(\delta) + \cos(\phi)\cos(\delta)\cos(h_{0)} = 0 \tag{2.33}$$

or

$$\cos(h_0) = -\tan(\phi)\tan(\delta) \tag{2.34}$$

Once $\tan(\varphi)\tan(\delta) > 1$, then the sun does not set and the sun is already risen at $h = \pi$, so $h_0 = \pi$. When $\tan(\varphi)\tan(\delta) < -1$, the sun does not rise and

$$Q^{-\text{day}} = 0.$$

$\dfrac{R_0^2}{R_E^2}$ is nearly constant over the course of a day and can be taken outside the integral

$$\int_{\pi}^{-\pi} Q dh = \int_{h_0}^{-h_0} Q dh = S_0\frac{R_0^2}{R_E^2}\int_{h_0}^{-h_0}\cos(\theta)dh$$

$$= S_0\frac{R_0^2}{R_E^2}\Big[h\sin(\phi)\sin(\delta) + \cos(\phi)\cos(\delta)\sin(h)\Big]_{h = h_0}^{h = -h_0}$$

$$= -2S_0\frac{R_0^2}{R_E^2}\Big[h_0\sin(\phi)\sin(\delta) + \cos(\phi)\cos(\delta)\sin(h_{0)}\Big] \tag{2.35}$$

Therefore:

$$Q^{-day} = \frac{S_0}{\pi}\frac{R_0^2}{R_E^2}\Big[h_0\sin(\phi)\sin(\delta) + \cos(\phi)\cos(\delta)\sin(h_{0)}\Big] \tag{2.36}$$

Since θ is considered the conventional polar angle describing a planetary orbit, $\theta = 0$ at the vernal equinox and the declination δ as a function of the orbital position would be:

$$\delta = \varepsilon \sin(\theta) \tag{2.37}$$

where ε is the obliquity and the conventional longitude of perihelion ϖ shall be related to the vernal equinox, so the elliptical orbit can be rewritten as:

$$R_E = \frac{R_0}{1 + e\cos(\theta - \omega)} \tag{2.38}$$

or

$$\frac{R_0}{R_E} = 1 + e\cos(\theta - \omega) \tag{2.39}$$

Here, ϖ, ε, and e are calculated from astrodynamical laws, so that a consensus of observations of Q^{-day} can be determined from any latitude φ and θ. However, $\theta = 0°$ is considered the duration of the vernal equinox, $\theta = 90°$ is exactly the time of the summer solstice, $\theta = 180°$ is exactly the time of the autumnal equinox and $\theta = 270°$ is exactly the time of the winter solstice. Therefore, the equation can be simplified for irradiance on a given day as follows:

$$Q = S_0 \left(1 + 0.034\cos\left(2\pi\frac{n}{365.25}\right)\right) \tag{2.40}$$

where n is the number of days of the year and thus the solar characteristics for both the theoretical function of optimum and modular to generate electricity can be shown per unit area (Figure 2.5).

Eventually, a peak high-frequency cutoff Ω_C of solar irradiance is calculated to keep the bifurcation of DOS away from the Earth's surface. Necessarily, a tipped high-frequency cutoff of the Earth's surface Ω_d is determined by controlling the positive DOS in 2D and 1D of the photon irradiance. Hence, $pi_2(x)$ acted as an algorithm function, and $e_{rfc}(x)$ acted as an additional function. Thus, the DOS of the Earth's surface, represented here as $\varrho_{PC}(\omega)$, is determined by calculating the photonic energy frequencies of Maxwell's rules into the Earth's surface. For 1D on Earth's surface, the represented DOS is thus expressed as $\varrho_{PC}(\omega) \propto \dfrac{1}{\sqrt{\omega - \omega_e}}\Theta(\omega - \omega_e)$,

where $\Theta(\omega - \omega_e)$ represents the Heaviside functional step and ω_e expresses the frequency of net solar energy generation (Figure 2.6).

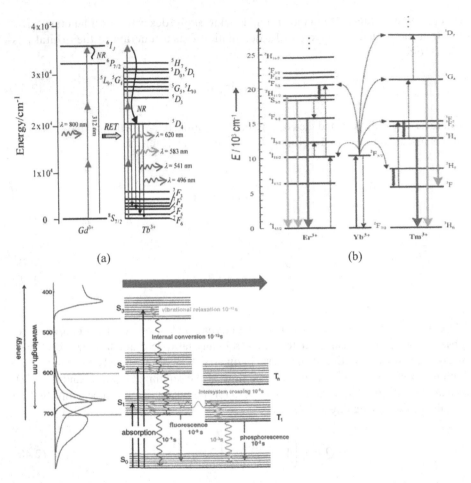

FIGURE 2.5 (a) shows the above stands for the solar energy function of the optimum working point for energy generation per unit area; (b) depicts the theoretical function of the modular point of energy generation per unit area; and (c) shows the energy generation at various wavelengths of the photon spectrum.

This DOS is thus determined to confirm a 3D isentropic function on the Earth's surface to acquire an accurate net qualitative state of the solar energy by inducing the non-Weisskopf–Winger mode of photons on the Earth's surface. Naturally, this 3D state will be the functional DOS in the PBE area of the DOS: $\varrho_{PC}(\omega) \propto \dfrac{1}{\sqrt{\omega - \omega_e}} \Theta(\omega - \omega_e)$,

and thus it has been integrated into the net electricity (EF) vector of the Earth's surface to accurately determine the net electricity energy generation on the Earth's surface (15). Considering both 2D and 1D, the photonic energy DOS is clarified by the pure algorithm of divergence, which is close to the PBE and thus expressed as

$\varrho_{PC}(\omega) \propto -\left[\ln\left|(\omega - \omega_0)/\omega_0\right| - 1\right]\Theta(\omega - \omega_e)$, where ω_e denotes the midpoint of the

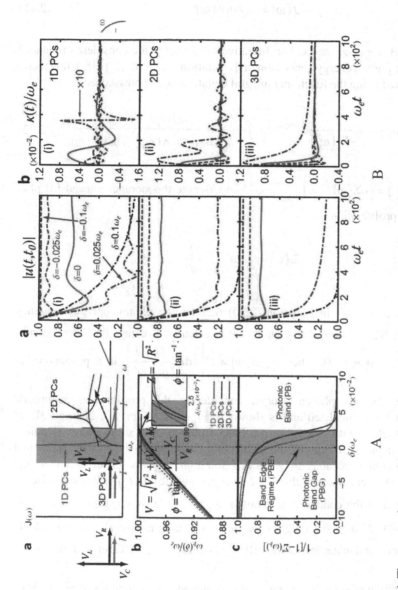

FIGURE 2.6 A The structural composition of photons and the rate of energy deliberation in the acting PV panel. (a) Functional area at different DOS magnitudes of 1D, 2D and 3D PV cells. (b) Photon frequency rate in the functional photonic band edge regime (PBE) and photonic band gap (PBG). (c) Photon's magnitude to deliver high energy into the functional photonic band edge regime (PBE) and photonic band gap (PBG). Figure 6B. Proliferation of photon dynamics into photovoltaic cells. (a) Considering the PB area, $<a(t)> = 5\, u(t, t_0)\, <a(t_0)>$, and (b) photon dynamic rate $k(t)$, functional variable for (i) 1D, (ii) 2D and (iii) 3D quantum fields into PV cells (Courtesy Figure 6A and 6B: Ping-Yuan Lo, Heng-Na Xiong & Wei-Min Zhang (2015); Scientific Reports, volume 5, Article number: 9423).

tip algorithm. The functional area $J(\omega)$ is thus clarified as the photon energy generation on the Earth's surface, where the solar energy generation of $V(\omega)$ depends on the total solar irradiance on the Earth's surface:

$$J(\omega) = \varrho(\omega)|V(\omega)|^2 \qquad (2.41)$$

Hence, the PB frequency ω_c and proliferative solar energy are considered functions $u(t,t_0)$ for the photon energy generation in the relation $a(t) = u(t,t_0)a(t_0)$. It is therefore determined using the functional integral equation and expressed as:

$$u(t,t_0) = \frac{1}{1 - \Sigma'(\omega_b)} e^{-i\omega(t-t_0)} + \int_{\omega_e}^{\infty} d\omega \frac{J(\omega)e^{-i\omega(t-t_0)}}{\left[\omega - \omega_c - \Delta(\omega)\right]^2 + \pi^2 J^2(\omega)} \qquad (2.42)$$

where $\Sigma'(\omega_b) = \left[\partial\Sigma(\omega)/\partial\omega\right]_{\omega=\omega_b}$ and $\Sigma(\omega)$ denote the storage-induced PB photonic energy proliferation:

$$\Sigma(\omega) = \int_{\omega_e}^{\infty} d\omega' \frac{J(\omega')}{\omega - \omega'} \qquad (2.43)$$

Here, the frequency ω_b in Equation (2.20) denotes the photon energy frequency module in PBG $(0 < \omega_b < \omega_e)$ and thus it is calculated using the areal condition: $\omega_b - \omega_c - \Delta(\omega_b) = 0$, where $\lesssim \Delta(\omega) = P\left[\int d\omega' \frac{J(\omega')}{\omega - \omega'}\right]$ is a primary-value integral.

Therefore, the net photon energy, considering the proliferation magnitude $|u(t,t_0)|$, has been calculated and is shown in Figure 2.6a for 1D, 2D, and 3D of the Earth's surface with respect to the PBG function. The solar energy dynamic rate $\kappa(t)$ is depicted in Figure 2.6b, neglecting the function $\delta = 0.1\omega_e$. The results revealed that emitted photons are generated at a high rate once ω_c crosses from the PBG to the PB area. Because the range in $u(t,t_0)$ is $1 \geq |u(t,t_0)| \geq 0$, the crossover area as related to the condition is denoted as $0.9 \gtrsim |u(t \to \infty, t_0)| \geq 0$, which corresponds to $-0.025\omega_e \lesssim \delta \lesssim 0.025\omega_e$, with a production rate $\kappa(t)$ within the PBG $(\delta < -0.025\omega_e)$ and in the area of the PBE $(-0.025\omega_e \lesssim \delta \lesssim 0.025\omega_e)$ of the Earth's surface.

The generation of solar energy emissions is almost exponential for $\delta \gg 0.025\omega_e$, which is a Markov factor. Figure 2.5a shows the dash-dotted black curves with $\delta = 0.1\omega_e$. In the crossover area $(-0.025\omega_e \lesssim \delta \lesssim 0.025\omega_e)$, the PB frequency of the PBE of the Earth's surface sharply increases the mode of photon energy generation emission (20,31). Thus, this proliferation of emitted solar photons confirms the net energy state photon in the Earth's surface of the PBG, where the photons are in a nonequilibrium photonic energy state (22,33).

Then, the solar irradiance on the entire Earth's surface is clarified considering thermal variation with respect to the solar energy concentration function $v(t,t)$ by determining the nonequilibrium solar energy scattering and reflecting calculations globally (11):

$$v(t,t) = \int_{t_0}^{t} dt_1 \int_{t_0}^{t} dt_2 u^*(t_1,t_0) \tilde{g}(t_1,t_2) u(t_2,t_0) \tag{2.44}$$

Here, the two-time correlation function of the Earth's surface $\tilde{g}(t_1,t_2) = \int d\omega J(\omega) \bar{n}(\omega,T) e^{-i\omega(t-t')}$ reveals the solar energy generation variations induced by the thermal relativistic condition of the Earth's surface, where $\bar{n}(\omega,T) = 1/[e^{\hbar\omega/k_B T} - 1]$ is the proliferation of the photon energy emission in the Earth's surface at the optimum temperature T and is expressed as:

$$v(t,t \to \infty) = \int_{\omega_e}^{\infty} d\omega \mathcal{V}(\omega) \text{ with}$$

$$\mathcal{V}(\omega) = \bar{n}(\omega,T) [\mathcal{D}_l(\omega) + \mathcal{D}_d(\omega)] \tag{2.45}$$

Here, Equation (2.21) is simplified to determine the nonequilibrium condition: $\mathcal{V}(\omega) = \bar{n}(\omega,T)\mathcal{D}_d(\omega)$. Under low-temperature conditions on the Earth's surface, Einstein's photon energy fluctuation dissipation is not dynamically viable at the PB on the Earth's surface but also connects the photonic energy state, which has been measured as the field intensity of the solar energy induction $n(t) = a^{\dagger}(t)a(t) = |u(t,t_0)|^2 n(t_0) v(t,t)$, where $n(t_0)$ represents the primary PB of the Earth's surface. Therefore, in Figure 2.6, the plotted net amount of photon energy versus temperature on the Earth's surface is clarified as the nonequilibrium proliferated photon energy generation, as shown by the solid-blue curve (Figure 2.6). To be more specific, the first PB of the Earth's surface has been considered as the Fock state photon number n_0, i.e. $\rho(t_0) = |n_0 n_0|$, which is obtained mathematically through the quantum dynamics of the photon energy and then by solving Equation (2.45), respecting the state of net photon energy production at time t:

$$\rho(t) = \sum_{n=0}^{\infty} \mathcal{P}_n^{(n_0)}(t) |n_0 n_0| \tag{2.46}$$

$$\mathcal{P}_n^{(n_0)}(t) = \frac{[v(t,t)]^n}{[1+v(t,t)]^{n+1}} [1 - \Omega(t)]^{n_0} \times \sum_{k=0}^{min\{n_0,n\}} \binom{n_0}{k}\binom{n}{k} \left[\frac{1}{v(t,t)}\frac{\Omega(t)}{1-\Omega(t)}\right]^k \tag{2.47}$$

where $\Omega(t) = \dfrac{|u(t,t_0)|^2}{1+v(t,t)}$. Therefore, the result reveals that the electron state photon energy will evolve into different Fock states of $|n_0 is\ \mathcal{P}_n^{(n_0)}(t)$ on the Earth's surface. The proliferation of the net photon energy dissipation $\mathcal{P}_n^{(n_0)}(t)$ in the primary state $|n_0 = 5$ and steady-state limit, $\mathcal{P}_n^{(n_0)}(t \to \infty)$ is thus shown in Figure 2.6. Therefore, the net photon energy generation on Earth's surface will ultimately reach the thermal nonequilibrium state, which is expressed as:

$$\mathcal{P}_n^{(n_0)}(t \to \infty) = \frac{\left[\bar{n}(\omega_c,T)\right]^n}{\left[1+\bar{n}(\omega_c,T)\right]^{n+1}} \tag{2.48}$$

To probe this enormous photon energy generation on the Earth's surface, a further calculation of the photon energy distribution within the quantum field of the Earth's surface has been conducted through the high-temperature coherent states and solving Equation (2.48), considering the energy state of photons, expressed by:

$$\rho(t) = \mathcal{D}\left[\alpha(t)\right]\rho_T\left[v(t,t)\right]\mathcal{D}^{-1}\left[\alpha(t)\right] \tag{2.49}$$

where $\mathcal{D}\left[\alpha(t)\right] = exp\left\{\alpha(t)\alpha^\dagger - \alpha^*(t)\alpha\right\}$ denotes the displacement functions with $\alpha(t) = u(t,t_0)\alpha_0$ and

$$\rho_T\left[v(t,t)\right] = \sum_{n=0}^{\infty} \frac{\left[v(t,t)^n\right]}{\left[1+v(t,t)\right]^{n+1}}|nn| \tag{2.50}$$

Here, ρ_T denotes a thermal state with an average particle quantum $v(t,t)$, where Equation (2.11) suggests that the peak point photon energy generation state will evolve into a thermal state (28,31), which is considered the functional state of the photon $\mathcal{D}\left[\alpha(t)\right]|n^{37}$ on the Earth's surface. Thus, the net photon energy generation calculation is represented by the following equation:

$$m|\rho(t)|n = J(\omega) = e^{-\Omega(t)|\alpha_0|^2}\frac{\left[\alpha(t)\right]^m\left[\alpha^*(t)\right]^n}{\left[1+v(t,t)\right]^{m+n+1}}$$

$$= \sum_{k=0}^{min\{m,n\}}\frac{\sqrt{m!n!}}{(m-k)!(n-k)!k!}\left[\frac{v(t,t)}{\Omega(t)|\alpha_0|^2}\right]^k \tag{2.51}$$

where the emission of the net photon energy ($m|\rho(t)|\ln$) into the Earth's surface, its conversion of photon energy into electricity $\left[1+v(t,t)\right]^{m+n+1}$ and the nonequilibrium condition $\left[\alpha(t)\right]^{m}\left[\alpha^{*}(t)\right]^{n}$ of the Earth's surface have been calculated.

2.3.2 MODELING NET ELECTRICITY ENERGY GENERATION FROM THE TOTAL SOLAR IRRADIANCE ON EARTH

To transform this tremendous amount of photon energy into electricity energy, the net solar energy is computed on a conceptual model of series and parallel circuits of the Earth's surface. The conceptual Earth surface is then hypothetically implemented into the *I–V* single diode circuit of the Earth's surface to obtain the precise *I–V* relationship of the net solar energy that reaches the Earth's surface by calculating the following equation:

$$I = IL - IO\left\{\exp\left[\frac{q(V+I_{RS})}{AkTc}\right] - 1\right\} - \frac{(V+I_{RS})}{R_{Sh}} \tag{2.52}$$

Here, I_L denotes the photon formation current, IO denotes the ideal current flow into the diode, R_s denotes the resistance in a series, A denotes the diode function, k (= 1.38 × 10^{-23} W/m²K) denotes Boltzmann's constant, q (=1.6 × 10^{-19}C) denotes the charge amplitude of the electron, and T_C denotes the Earth's temperature. Consequently, the *I–q* linked on the Earth's surface varies into the diode cell, which is expressed as the dynamic current as follows (25):

$$I_O = I_{RS}\left(\frac{T_C}{T_{ref}}\right)^3 \exp\left[\frac{qEG\left(\frac{1}{T_{ref}} - \frac{1}{T_C}\right)}{KA}\right] \tag{2.53}$$

where I_{RS} denotes the dynamic current representing the functional transformation of solar radiation and qEG denotes the band-gap solar radiation into the conceptual Earth surface at different DOS dimensional modes of 1D, 2D, and 3D (Table 2.1).

Here, considering this conceptual Earth surface, with the exception of the *I–V* curve, a calculative result of the linked *I–V* curves among all of the conceptual solar cells has been determined (11.26). Thus, the equation is rewritten as follows to determine the V-R relationship much more accurately:

$$V = -IR_s + K\log\left[\frac{I_L - I + I_O}{I_O}\right] \tag{2.54}$$

TABLE 2.1
Solar Energy in Various DOS Dimensional Modes Reaches the Earth's Surface (ES) with Various Corresponding Unit Areas $J(\omega)$ and Self-Energy-induced Reservoirs of $\Sigma(\omega)$.

Solar Energy (ES)	Unit area $J(\omega)$ for different DOS	Solar energy correction in the Earth's surface $\Sigma(\omega)$
1D	$\dfrac{ES}{2\pi r}\dfrac{1}{\sqrt{\omega-\omega_e}}\Theta(\omega-\omega_e)$	$-\dfrac{ES}{\sqrt{2\omega_e-\omega}}$
2D	$-ES\left[\ln\left\|\dfrac{\omega-\omega_0}{2\omega_0}\right\|-1\right]$ $\Theta(\omega-\omega_e)\Theta(\Omega_d-\omega)$	$ES\left[Li_2\left(\dfrac{\Omega_d-\omega_0}{\omega-\omega_0}\right)-Li_2\left(\dfrac{\omega_0-\omega_e}{\omega_0-\omega}\right)\right.$ $\left.-\ln\dfrac{\omega_0-\omega_e}{\Omega_d-\omega_0}\ln\dfrac{\omega_e-\omega}{\omega_0-\omega}\right]$
3D	$ES\sqrt{\dfrac{2\omega-\omega_e}{\Omega_C}}\exp\left(-\dfrac{\omega-\omega_e}{\Omega_C}\right)$ $\Theta(\omega-\omega_e)$	$ES\left[\pi\sqrt{\dfrac{\omega_e-\omega}{\Omega_C}}\exp\left(-\dfrac{2\omega-\omega_e}{\Omega_C}\right)\right.$ $\left.erfc\sqrt{\dfrac{\omega_e-\omega}{\Omega_C}}-\sqrt{2\pi r}\right]$

The functional C, η and χ act as coupled forces between the solar energy at the Earth's surface of 1D, 2D, and 3D areal surfaces (Courtesy: Ping-Yuan Lo, Heng-Na Xiong & Wei-Min Zhang (2015). *Scientific Reports*, volume 5, Article number: 9423)

where K denotes the constant $\left(=\dfrac{AkT}{q}\right)$ and I_{mo} and V_{mo} are denoted as the net current and voltage in the conceptual Earth surface, respectively. Subsequently, the relationship among I_{mo} and V_{mo} remains motional in the I–V Earth surface, which can be written as:

$$V_{mo} = -I_{mo}R_{Smo} + K_{mo}\log\left(\frac{I_{Lmo}-I_{mo}+I_{Omo}}{I_{Omo}}\right) \tag{2.55}$$

where I_{Lmo} denotes the photon-induced current, I_{Omo} denotes the dynamic current into the diode, R_{smo} denotes the resistance in series, and K_{mo} denotes the factorial constant.

Once all nonseries (NS) cells are interlinked in the series, then the series resistance is calculated as the sum of each solar cell series resistance $R_{smo} = N_S \times R_S$ current considering the functional coefficient of the constant factor $K_{mo} = N_S \times K$. The flow of current dynamics into the circuit is linked to the cells in a series connection (2). Thus, the current dynamics in Equation (2.41) remain the same in each part of $I_{omo} = I_o$ and $I_{Lmo} = I_L$. Thus, the mode of the $I_{mo}-V_{mo}$ relationship for the N_S series of connected cells can be expressed by:

$$V_{mo} = -I_{mo} N_S R_S + N_S K \log\left(\frac{I_L - I_{mo} + I_o}{I_o}\right)$$ (2.56)

Naturally, the current–voltage relationship can be further modified considering all parallel links into the N_p cell connections in all parallel modes and can be described as follows (25,27):

$$V_{mo} = -I_{mo}\frac{Rs}{Np} + K \log\left(\frac{N_{sh} I_L - I_{mo} + N_p I_o}{N_p I_o}\right)$$ (2.57)

Since the photon-induced current primarily depends on the solar radiation and optimum temperature configuration, the net current dynamic is calculated as:

$$I_L = G\left[I_{SC} + K_I\left(T_C - T_{ref}\right)\right] * V_{mo}$$ (2.58)

where I_{sc} denotes the current at 25°C and KW/m², K_I denotes the Earth's surface coefficient factor, T_{ref} denotes the optimum temperature, and G denotes the solar energy in mW/m².

Finally, the electricity energy generation around the Earth's surface has been computed to confirm the net emitted photon utilization by integrating the local Albanian electric fields; thus, the global U (1) gauge field will allow for the addition of a mass term of the functional particle of $\varnothing' \rightarrow e^{i\alpha(x)}\varnothing$. It is then further clarified by explaining the variable derivative of the transformation law of the scalar field using the following equation (30,31):

$$\partial_\mu \rightarrow D_\mu = \partial_\mu = ieA_\mu\left[\text{covariant derivatives}\right]$$

$$A'_\mu = A_\mu + \frac{1}{e}\partial_\mu \alpha[A_\mu \text{ derivatives}]$$ (2.59)

Here, the global U (1) gauge denotes the invariant local Albanian for a complex scalar field, which can be further expressed as:

$$\mathcal{G} = \left(D^\mu\right)^\dagger \left(D_\mu \varnothing\right) - \frac{1}{4}F_{\mu\nu}F^{\mu\nu} - V(\varnothing)$$ (2.60)

The term $\dfrac{1}{4} F_{\mu\nu} F^{\mu\nu}$ is the dynamic term for the gauge field of the Earth's surface, and $V(\varnothing)$ denotes the extra term in the local Albanian, which is $V(\varnothing^*\varnothing) = \mu^2 (\varnothing^*\varnothing) + \lambda (\varnothing^*\varnothing)^2$.

Therefore, the generation of local Albanian ($\mathcal{h}$) under the perturbational function of the quantum field of the Earth's surface has been confirmed by the calculation of the mass-scalar particles ϕ_1 and ϕ_2 along with a mass variable of μ. Under this condition, $\mu^2 < 0$ had an infinite number of quantum, which is clarified by $\phi_1^2 + \phi_2^2 = -\mu^2 / \lambda = v^2$, and $\mathcal{h}$ through the variable derivatives using further shifted fields η *and* ξ defined the quantum field as $\phi_0 = \dfrac{1}{\sqrt{2}} \left[(v + \eta) + i\xi \right]$.

Kinetic term: $\mathcal{h}\,(\eta,\,\xi) = (D^\mu\phi)^*(D^\mu\phi)$

$$= \left(\partial^\mu + ieA^\mu \right) \phi^* \left(\partial_\mu - ieA_\mu \right) \phi \tag{2.61}$$

Thus, this expanding term in $\mathcal{h}$ associated with the scalar field of the Earth's surface suggests that the net Earth's surface field is prepared to initiate net electricity energy generation into its quantum field of induced photon energy (Figure 2.7).

To determine this electricity energy, hereby, a nonvariable function of readily available dynamics has been implemented for the calculation of $\overline{\phi}\,[s_0]$ to confirm the expected value of s_0 considering the Earth's surface, which is expressed as follows:

$$\overline{\phi}\left[s_0\right] = 2s_0 \left(ln4s_0 - 2 \right) + ln4s_0 \left(ln4s_0 - 2 \right) - \frac{(\pi^2 - 9)}{3}$$
$$+ s_0^{-1} \left(ln4s_0 + \frac{9}{8} \right) + \ldots \left(s_0 \gg 1 \right); \tag{2.62}$$

$$\overline{\phi}\left[s_0\right] = \left(\frac{2}{3} \right) (S_0 - 1)^{\frac{3}{2}} + \left(\frac{5}{3} \right) (S_0 - 1)^{\frac{5}{2}} - \left(\frac{1507}{420} \right)$$
$$(S_0 - 1)^{\frac{7}{2}} \ (1/2 \text{ instead of } 1). \tag{2.63}$$

Then, the final equation can be rewritten as follows, where s_0 is the areal value of electricity energy generation into the Earth's surface ($1 \ m^2$):

$$\overline{\phi}\left[s_0\right] = \left(\frac{2}{3} \right) (S_0 - 1)^{\frac{3}{2}} + \left(\frac{5}{3} \right) (S_0 - 1)^{\frac{5}{2}} - \left(\frac{1507}{420} \right) (S_0 - 1)^{\frac{7}{2}} \tag{2.64}$$

The function $\overline{\phi}\,[s_0]$ thus determines the net electricity energy generation from the total solar energy into the atmosphere by calculating Earth's cross-sectional area of 127,400,000 km², and the net solar energy intercepted by the surface of the Earth is 1.740×10^{17} Watts. Considering seasonal and climate variations, the net power

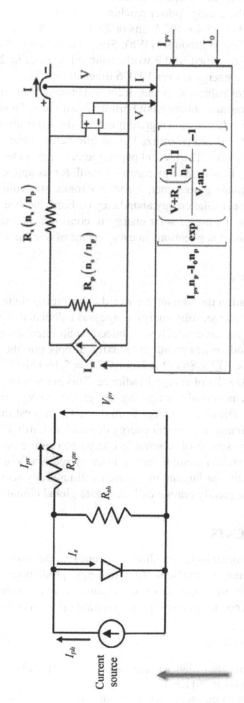

FIGURE 2.7 (a) The scalar field of the Earth's surface; (b) solar energy scalar field on Earth's surface; and (c) net electricity current energy generation on Earth from the total solar energy on Earth (12, 21).

reaching the ground generally averages at 200 Watts per square meter per day (2, 33). Thus, at any time the average power reaching the Earth's surface is calculated as $127.4 \times 106 \times 106 \times 200 = 25.4 \times 10^{15}$ Watts or 25,400 Tera Watts, which is TW $\times$ 24 $\times$ 365 = 222,504,000 Tera Watthours (TWh). Since the net annual electrical energy (not the total energy) consumed in the world from all sources in 2019 was 22,126 TWh, the available solar energy is over 10,056 times the world's consumption.

Global environmental vulnerability has become a crucial issue due to greenhouse gas concentrations in the atmosphere from burning fossil fuels. Thus, the use of solar energy over the past decade has shown great promise, since it is abundantly available anywhere in the world and can be utilized as an alternative clean energy source in every sector of our daily lives. The present paper, therefore, provides a detailed calculation of the abundancy of solar energy reaching Earth for its application as a source of clean energy to mitigate global energy and environmental vulnerability. Simply put, in this research the net solar energy abundancy on Earth has been mathematically calculated to determine the use of solar energy to confirm a realistic assessment of this energy for commercial applications in every sector of our daily lives.

2.4 CONCLUSION

Since fossil fuel utilization throughout the world is becoming finite and is the major contributor to climate change, solar energy usage was calculated in this research as a renewable clean energy source, which will indeed be an interesting source for mitigating global energy and environmental perplexity. Simply put, the solar energy that reaches Earth is $1.73 \times 10^{17} \times 86400 \times 365.2422 \sim= 5.46 \times 10^{24}$ J annually, which is equivalent to 5,460,000 EJ of energy irradiance. This amount of energy can surely play a vital role in tremendously mitigating the global energy crisis and reducing greenhouse emissions. The radiant energy from the sun is thus estimated to be 10,056 times higher than the annual net current energy demand on Earth; therefore, it should be used as the primary source of renewable energy to capture and mitigate global energy needs. The utilization of solar energy in every sector of daily life will surely have tremendous benefits for humankind, which will naturally secure global energy consumption and subsequently dramatically mitigate global climate change.

ACKNOWLEDGMENTS

This research was supported by Green Globe Technology, Inc. under grant RD-02020-04 for building a better environment. Any findings, predictions, and conclusions described in this article are solely those of the authors, who confirm that the article has no conflicts of interest for publication in a suitable journal or book chapter.

REFERENCES

1.   A. Reinhard. Strongly correlated photons on a chip. Nat. Photonics 6, 2–4 (2011). DOI: 10.1038/nphoton.2011.321
2.   C. Lei, U. Zhang. A quantum photonic dissipative transport theory. Ann. Phys. 327, 1408–1412 (2012).

3. C. Sayrin. Real-time quantum feedback prepares and stabilizes photon number states. Nature 477, 73–76 (2011).

4. C. Song. Fano resonance analysis in a pair of semiconductor quantum dots coupling to a metal nanowire. Opt. Lett. 37, 978–980 (2012).

5. C. Wang, C. Zhang, R. Xiao, J. Multiple plasmon-induced transparencies in coupled-resonator systems. Opt. Lett. 37, 5133–5135 (2012).

6. D. Englund. Resonant excitation of a quantum dot strongly coupled to a photonic crystal nanocavity. Phys. Rev. Lett. 104, 904–909 (2010).

7. D. O'Shea, C. Junge. Fiber-optical switch controlled by a single atom. Phys. Rev. Lett. 111, 193–201 (2013).

8. D. Roy. Two-photon scattering of a tightly focused weak light beam from a small atomic ensemble: An optical probe to detect atomic level structures. Phys. Rev. A. 87, 638–6419 (2013).

9. E. Saloux. Explicit model of photovoltaic panels to determine voltages and currents at the maximum power point. Sol. Energy 66, 450–459 (2015).

10. G. Yan, H. Lu, A. Chen. Single-photon router: Implementation of Information-Holding of Quantum States. Int. J. Theor. Phys. 24, 78–83 (2016).

11. H. Faruque. Transforming dark photons into sustainable energy. Int. J. Energy Environ. Eng. 11, 38–45 (2018).

12. H. Faruque. Green science: Advanced building design technology to mitigate energy and environment. Renew Sustain Energy Rev. 56, 87–93 (2018).

13. H. Faruque. Sustainable technology for energy and environmental benign building design. J. Build. Eng. 44, 123–132 (2019).

14. H. Faruque. Photon energy amplification for the design of a micro PV panel. Int. J, Energy Res. 66, 340–349 (2018).

15. H. Faruque. Green science: Independent building technology to mitigate energy, environment, and climate change. Renew Sustain Energy Rev. 45, 223–229 (2017).

16. H. Faruque. Design and construction of ultra-relativistic collision PV panel and its application into building sector to mitigate total energy demand. J. Build. Eng. 36, 270–278 (2017).

17. H. Faruque. Solar energy integration into advanced building design for meeting energy demand and environment problem. Int. J. Energy Res. 78, 671–679 (2016).

18. H. Min, C. Veronis. Subwavelength slow-light waveguides based on a plasmonic analogue of electromagnetically induced transparency. Appl. Phys. Lett. 99, 143–151 (2011).

19. J. Douglas, H. Habibian, C. Hung, A. Gorshkov, H. Kimble, D. Chang. Quantum many-body models with cold atoms coupled to photonic crystals. Nat. Photonics 23, 45–49 (2015).

20. J. Hou, H. Wang, P. Liu. Applying the blockchain technology to promote the development of distributed photovoltaic in China. Int. J. Energy Res. 29, 450–459 (2018).

21. J. Huang, F. Shi, T. Sun. Controlling single-photon transport in waveguides with finite cross section. Phys. Rev. A. 88, 13836–13842 (2013).

22. J. Liao, Q. Law. Correlated two-photon transport in a one-dimensional waveguide side-coupled to a nonlinear cavity. Phys. Rev. A. 82, 538–545 (2010).

23. J.S. Douglas, H. Habibian, C.-L. Hung, A.V. Gorshkov, H.J. Kimble, D.E. Chang. Quantum many-body models with cold atoms coupled to photonic crystals. Nat. Photonics 9, 326–331 (2015).

24. J. Park, H. Kim, Y. Cho, C. Shin. Simple modeling and simulation of photovoltaic panels using Matlab/Simulink. Adv. Sci. Technol. Lett. 73, 147–155 (2014).

25. J. Eichler, T. Stöhlker. Radiative electron capture in relativistic ion-atom collisions and the photoelectric effect in hydrogen-like high-Z systems. Phys. Rep. 439, 1–99 (2007).

26. J.J. Soon, K.S. Low. Optimizing photovoltaic model parameters for simulation. IEEE Int. Symp. Ind. Electron. 1813–1818 (2012).

27. J.T. EichlerStöhlker. Radiative electron capture in relativistic ion-atom collisions and the photoelectric effect in hydrogen-like high-Z systems. Phys. Rep. 439, 1–99 (2007).

28. K.G. Sharma, A. Bhargava, K. Gajrani. Stability analysis of DFIG based wind turbines connected to electric grid. IREMOS 6, 879–887 (2013).

29. K. Oulton, R. Zhang. Nonlinear quantum optics in a waveguide: Distinct single photons strongly interacting at the single atom level. Phys. Rev. Lett. 106, 113–119 (2011).

30. L. Langer, S.V. Poltavtsev, I.A. Yugova, M. Salewski, D.R. Yakovlev, G. Karczewski, T. Wojtowicz, I.A. Akimov, M. Bayer. Access to long-term optical memories using photon echoes retrieved from semiconductor spins. Nat. Photonics 8, 851–857 (2014).

31. L. Yang, S. Wang, Q. Zeng, Z. Zhang, T. Pei, Y. Li, L.M. Peng. Efficient photovoltage multiplication in carbon nanotubes. Nat. Photonics 5, 672–676 (2011).

32. M.C. Güçlü, J. Li, A.S. Umar, D.J. Ernst, M.R. Strayer. Electromagnetic lepton pair production in relativistic heavy-ion collisions. Ann. Phys. 272, 7–48 (1999).

33. M. Reed, L. Maxwell. Connections between groundwater flow and transpiration partitioning. Science 353, 377–380 (2015).

34. P. Longo, P. Schmitteckert. Few-photon transport in low-dimensional systems. Phys. Rev. A. 83, 638–647 (2011).

35. T. Pregnolato, E. Song. Single-photon non-linear optics with a quantum dot in a waveguide. Nat. Commun. 6, 86–95 (2015).

36. T. Shi, S. Fan. Two-photon transport in a waveguide coupled to a cavity in a two-level system. Phys. Rev. A. 84, 638–643 (2011).

37. Y. Xiao, C. Meng, P. Wang, Y. Ye. Single-nanowire single-mode laser. Nano Lett. 48, 340–348 (2011).

38. Y. Wang, Y. Zhang, Q. Zhang, B. Zou, U. Schwingenschlogl. Dynamics of single photon transport in a one-dimensional waveguide two-point coupled with a Jaynes-Cummings system. Sci. Rep. 98, 451–458 (2016).

3 Energy Storage
A Practical Solution to Increase Wind Energy Integration in the US Electricity Sector

Marissa Schmauch Womble and George Xydis

3.1 INTRODUCTION

In the United States, about 2.4 percent of primary energy comes from wind power (U.S. Energy Information Administration, 2019a). In the electricity sector, about 7.3 percent of total utility scale electricity generation is derived from wind (U.S. EIA, 2020). Although these figures have increased in recent years, the current installed wind capacity, ~98,000 MW, is less than 1 percent of the potential wind capacity based on wind resource (U.S. Department of Energy, n.d.). Two key characteristics of wind power have precluded wind from achieving greater penetration. First, wind power is variable on both short and long timescales (Xydis, Pechlivanoglou, & Nayeri, 2015). Wind speeds vary across the country, and vary between times of day, days, seasons, and years. Because there is variability in wind speed, there is also variability in the power output of wind farms. Second, unlike conventional generation, wind power is not easily dispatchable. A wind farm's power output cannot be increased to meet a dispatch request. These two factors, combined with a host of other issues including existing fossil fuel infrastructure, the economic competitiveness of new wind installations, and "not in my backyard" (NIMBY) opposition, have led to actual wind generation that is much lower than potential wind generation (Enevoldsen et al., 2019). Energy storage has the potential to mitigate both of these concerns and achieve greater wind penetration in the US electricity sector, a critical step toward decarbonizing the electricity sector and achieving a global transition to 100 percent renewable energy.

DOI: 10.1201/9781003240129-3

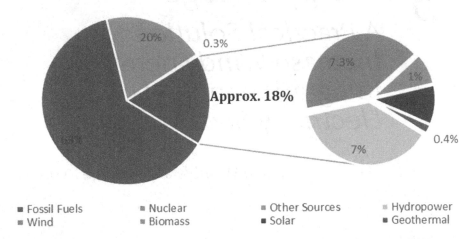

FIGURE 3.1   US utility-scale electricity generation by energy source (U.S. EIA, 2020).

Energy storage technologies, both old technologies like pumped storage hydropower (PSH) and more cutting-edge technologies like green hydrogen and battery storage (Michalitsakos, Mihet-Popa, & Xydis, 2017), have the potential to be deployed in combination with renewable generation resources and displace existing fossil fuel infrastructure. Energy storage paired with wind can also eliminate the curtailment of these valuable renewable energy resources (Ucal & Xydis, 2020). By combining energy storage with wind generation, wind energy penetration in the US electricity sector can be greatly increased. This has the effect of lowering carbon emissions associated with electricity generation and paving the way for a higher penetration of renewable energy on the US grid (Figure 3.1).

3.2   LITERATURE REVIEW

Some of the barriers to increased penetration of wind energy are the resource variability both geographically and in time. The potential for wind power generation is not evenly spread across the United States. Specifically, the areas with high wind potential are located in the middle of the country: Texas, Oklahoma, Kansas, Nebraska, Wyoming, Iowa, North Dakota, and South Dakota. The five top wind power producing states are Texas, Oklahoma, Iowa, Kansas, and California (U.S. Energy Information Administration, 2019b). California has a high population density, but the rest of these states are far away from the population centers on the coasts. To mitigate the geographic variability, the most appealing solution is using high-voltage direct current (HVDC) power lines to connect areas of high supply and high demand. HVDC minimizes transmission losses and can carry power with lower losses compared to alternating current (AC) lines at similar voltages. However, barriers to HVDC infrastructure include very high capital costs and difficulty reaching a consensus among the jurisdictions that will be impacted by the HVDC project (proposed HVDC projects span several states) (Merchant, 2019). Furthermore, the geographic variability of wind resources may be mitigated through the use of offshore wind that does have high resource potential near the population centers, or using other renewable energy resources that are more abundant on the coasts, such as solar power, biomass, or hydropower.

3.2.1 ENERGY DEMAND, VARIABILITY, AND DEMAND-BASED SOLUTIONS

The variability of wind in time occurs on daily, seasonal, and even longer timescales. Unlike a coal or natural gas power plants, which can be turned on and quickly (30 minutes or less (Peake, 2018) for coal and about 23 minutes for simple-cycle natural gas (Gonzales-Salazar, Kirsten, & Prchlik, 2018) heated up to 100 percent load), wind turbines vary in their power output and cannot be turned on or turned up on demand. The variability is not always predictable. On a daily basis, the power output varies in a relatively narrow range and is typically predictable. However, the predictability is increasing with shortened time horizon, but the variability still exists. Therefore, in electricity markets, intra-day markets are expected to expand as they can better accommodate that variability than the day-ahead markets (Wang et al., 2019).

Wind power output also varies on a seasonal basis, in part due to changes in weather patterns, and in part due to air density variation between seasons (Xydis, 2012). These changes in air density can lead to a change in capacity factor as great as ±6% (Ulazia, Saenz, Ibarra-Berastegi, Gonzales-Roji, & Carreno-Madiabeitia, 2019). The differences in weather between seasons that cause changes in wind speed are also well understood and therefore able to be forecasted with high accuracy (Soret et al., 2019). For example, in the northwestern United Kingdom, ocean cyclones and the associated winds are particularly strong in fall, winter, and spring (Peake, 2018). On longer timescales, prediction is not accurate. From one year to another, variations in total energy can be as great as nearly 40 percent due to long-term climate and weather patterns (Wan, 2012).

Electricity demand also varies throughout the course of days and seasons. The electricity demand depends upon weather and consumer behavior. Electricity consumption usually goes through a predictable daily cycle with peak hours between 7:00 a.m. and 11:00 p.m. on weekdays, in most cases with two peaks during the day. On a seasonal basis, more power is typically consumed in summer and winter than spring and fall due to increased space heating and cooling requirements (U.S. EIA, 2020). The daily and seasonal electricity demand vary due to anthropogenic factors like cooking times throughout the day or a desire to reduce one's energy consumption by lowering home temperature during the winter and non-anthropogenic factors like weather. Grid operators use many different modeling techniques to predict and then schedule electric power or heating power demand (Karabiber & Xydis, 2020). In the United States, grid reliability is high due to a redundant system, accurate predictions, and large generation reserves. When electricity demand is higher than supply, some reserves must be brought online to match the supply. When electricity demand is lower than supply, some generation sources must be taken offline to match the supply.

   Some of this matching can be handled via demand response (DR) and curtailment. Although curtailment is being used as one of the tools that transmission system operators (TSOs) have at their disposal, in practice the more the TSOs are using it, the less renewable energy is penetrated and significant amounts of wind/solar production are thrown away. DR refers to lowering the electricity demand to match supply during times when demand is peaking and is higher than electricity supply. DR may be accomplished by using smart grids, time-of-use (ToU) tariffs, or agreements by consumers (via incentives) to decrease electricity consumption (Andrews & Jelley, 2017). It is not so easy everywhere though. In many countries, especially where electricity markets are still regulated, there are no clear economic incentives for consumers to respond to hourly changes in electricity market prices, as a result, they might pay higher prices during a peak period than off period—and this could increase also the country's electricity production costs. It is generally the modification of consumers' response in energy demand through different methods such as financial incentives and consumers' training that offers the opportunity to the grid for general adjustments when needed.

   In the electricity sector, a useful tool is also demand side management (DSM). DSM is the mean to reduce peak electricity demand so that utilities can delay building further capacity. By reducing the overall load on an electricity network, DSM provides several beneficial effects such as mitigating electrical system emergencies, reducing the number of blackouts, and increasing system reliability. Other benefits can be reducing dependency on expensive imports of fuel, reducing energy prices, and reducing the effects on the environment by reducing harmful emissions. Thus, DSM applied to electricity systems provides significant economic, reliability, and environmental benefits. The major goal of DSM is to "persuade" consumers to use less energy during peak hours or to change the time of energy use to the off-peak hours (Panagiotidis, Effraimis, & Xydis, 2019). Therefore, applying DSM as a form of standing reserve could improve the system performance by increasing the amount of wind energy that can be absorbed as fewer generating units are scheduled to operate, which is particularly relevant when there is high production of wind energy with low demand. In this case, DSM would allow a bigger amount of wind to be absorbed and would therefore reduce the fuel burned. The value of storage and DSM when providing standing reserves can be determined by evaluating the improvements in the performance of the system (fuel cost and CO_2 emissions) (Strbac, 2008).

   Curtailment is "a reduction in the output of a generator from what it could otherwise produce given available resources (e.g., wind or sunlight), typically on an involuntary basis" (Bird, Cochran, & Wang, 2014). Lowering electricity supply to match demand is only one of the reasons for curtailment. Curtailment can also be necessitated by transmission congestion, lack of transmission access, issues with electricity voltage matching, issues with interconnections, and issues with electricity frequency matching (Bird, Cochran, & Wang, 2014). Though its result is undesirable, in practice the magnitude of curtailment in the United States is currently relatively small; in 2018 average capacity factor reduction as a result of curtailment was 0.7 percent. In some cases, curtailment may rise as wind penetration increases, but this is not the case in each independent system operator (ISO), and the rates of curtailment will not rise uniformly across all ISOs (U.S. Department of Energy, 2019). Instead of lowering supply when the price of fuel is free, essentially throwing away the potential to generate electricity from a source that is both economically free and free of carbon emissions, a more attractive option is to store the energy. This allows use at a later time when demand is higher and/or electricity production from wind and other sources is lower.

3.2.2 ENERGY STORAGE SOLUTIONS

The intermittency of wind energy and the technical capabilities of grids limit wind penetration. Energy storage technologies are obviously the proposed solution. For instance, compressed air energy storage (CAES) is a bulk energy storage technology, which may recover wind curtailments and benefit from the introduction of natural gas. CAES are usually based on large scale—even on national level—grid-connected applications used to supplement base load power stations. On the other hand, pumped-hydro storage systems, which are usually placed at isolated regions and are able to make use of the rejected wind energy amounts produced by local wind farms, seem to gain interest over the world and become essential as regard to higher shares of renewable generated electricity.

Batteries are one technology capable of storing renewable electricity. Batteries are a maturing technology, with 11 GWh globally installed (IRENA, 2019), and 708 MW capacity installed in the United States (Laporte, 2019). The capacity of utility-scale battery storage is projected to continue growing. Batteries have been demonstrated to help with the integration of renewable resources, including wind, into the electricity mix. Most of the existing utility-scale battery installations (90 percent) are lithium-ion (Li-ion) type (International Energy Agency, 2019). The storage duration of Li-ion batteries is typically around 4 hours (Laporte, 2019), which is well suited for smoothing the daily electricity supply curve from wind. Batteries are technically capable of meeting dispatch requests for peak demand, with instantaneous response time (IRENA, 2019). Since battery installations may be built anywhere, pairing batteries with wind is a very accessible solution for smoothing electricity supply variability. However, the capital cost of battery storage is high, $269/kWh (Mongird et al., 2019). Another disadvantage is presented by environmental concerns associated with lithium. Lithium mining is extremely water-intensive, with one ton of lithium requiring 1,900 tons of water to extract. Lithium ion battery recycling technology is nascent, and advancements in battery disassembly and recovery technology are likely to improve (Harper et al., 2019). Only about 5 percent of lithium-ion batteries are currently recycled in the European Union (EU) and the United States. There is no clear path to large-scale recycling of these batteries, and today researchers and manufacturers typically do not optimize the battery design to facilitate easy disassembly and recycling (Jacoby, 2019). Lithium is a finite resource with about 17 million tons in known reserves and 80 million tons in identified resources, which are currently undiscovered but are expected to be discovered in the future. In the United States, between 2,000 and 3,000 tons of lithium are consumed each year, with approximately 65 percent of this used in batteries. At a global consumption rate of 57,700 tons per year (the usage in 2019), the known reserves will last about 195 years (USGS, 2020). However, these consumption rates are in flux due to increased use of lithium-ion batteries in electric vehicles, battery storage, and electronics, and could increase, leading to shorter-lived reserves. Continued work to develop lithium-ion battery recycling techniques is imperative to the continued growth of lithium-ion batteries in electric vehicles, utility-scale battery installations, and other applications.

PSH is another promising technology. The main advantage of this technology is that it is readily available. It uses the power of water, a highly concentrated renewable energy source. This technology is currently the most used for high-power applications (a few tens of GWh or 100 of MW). Pumped storage subtransmission stations will be essential for the storage of electrical energy. The principle is generally well known: during periods when demand is low, these stations use electricity to pump the water from the lower reservoir to the upper reservoir. When demand is very high, the water flows out of the upper reservoir and activates the turbines to generate high-value electricity for peak hours. PSH systems have a conversion efficiency, from the point of view of a power network, of about 65–80 percent, depending on equipment characteristics (Menéndez & Fernández-Oro, 2020).

PSH is a mature technology with 153 GW installed globally (IRENA, 2017) and 22.9 GW installed in the United States (Mayes, 2019). PSH has a longer storage duration than batteries, as evaporation of some surface water will only result in minor losses of the stored gravitational potential energy. This makes PSH well suited for smoothing the seasonal variability in electricity supply from wind power. PSH is also technically capable of meeting dispatch requests for peak demand, with a response time in the order of 1–5 minutes (IRENA, 2017). However, there are disadvantages of PSH that hinder its widespread adoption and integration into the energy storage portfolio. First are high capital costs, about \$165/kWh (Mongird et al., 2019). Additionally, PSH installations are highly location constrained, as they can only be built where the local geography is suitable—either a high elevation drop on a river where a dam with a reservoir can be built, or existing bodies of water with an elevation drop between them. Because PSH installations can have a massive impact on waterways and the local environment, the permitting process for these installations is lengthy, which may hinder project development steps and developers' desire to make a large initial investment in PSH.

There are also a number of developing energy storage technologies that hold promise for energy storage to increase penetration of wind energy and other variable renewable electricity sources. The most well-developed technologies are hydrogen storage and compressed air energy storage. Hydrogen energy storage has a large global demand, about 8 exajoules (Hydrogen Council, 2017). However, most of the hydrogen that meets this demand is not green hydrogen (hydrogen that comes from renewable energy sources); approximately 95 percent comes from fossil fuels (IRENA, 2017). Lately, even small-scale applications were found to be meaningful, especially for the urban environment (Apostolou, Casero, Gil-Hernández, & Xydis, 2021). Hydrogen energy storage uses fuel cells. Fuel cells are a means of restoring spent energy to produce hydrogen through water electrolysis. The storage system proposed includes three key components: electrolysis, which consumes off-peak electricity to produce hydrogen; the fuel cell, which uses that hydrogen and oxygen from air to generate peak-hour electricity; and a hydrogen buffer tank to ensure adequate resources in periods of need. Oxidation–reduction between hydrogen and oxygen is a particularly simple reaction, which occurs within a structure (elementary electro-chemical cell) made up of two electrodes (anode–cathode) separated by electrolyte, a medium for the transfer of charge as ions. Fuel cells can be used in decentralized pro-duction (particularly low-power stations residential, emergency, etc.), spontaneous supply related or not to the network, mid-power cogeneration (a few 100 kW), and centralized electricity production without heat upgrading. They can also represent a solution for isolated areas where the installation of power lines is too difficult or expensive (mountain locations and so on).

CAES is a new technology that is also capable of providing electricity to meet peak demand, with a response time of 3–10 minutes (Mongird et al., 2019). There are two compressed air energy storage installations globally—one in the United States and one in Germany (IRENA, 2017). Compressed air has a high capital cost, although lower than PSH or lithium-ion batteries: $105/kWh (Mongird et al., 2019). CAES relies on relatively mature technology with several high-power projects in place. A power plant with a standard gas turbine uses nearly two-thirds of the available power to compress the combustion air. It therefore seems possible, by separating the processes in time, to use electrical power during off-peak hours (storage hours) in order to compress the air, and then to produce, during peak hours (retrieval hours), three times the power for the same fuel consumption by expanding the air in a com-bustion chamber before feeding it into the turbines. Residual heat from the smoke is recovered and used to heat the air. Compressed air energy storage is achieved at high pressures (40–70 bars) at near ambient temperatures. This means less volume and a smaller storage reservoir. Large caverns made of high-quality rock deep in the ground, ancient salt mines, or underground natural gas storage caves are the best options for compressed air storage, as they benefit from geostatic pressure, which facilitates the containment of the air mass. A large number of studies have shown that the air could be compressed and stored in underground, high-pressure piping (20–100 bars). This method would eliminate the geological criteria and make the system easier to operate (Zhang et al., 2015).

Some technologies that are currently in the prototyping phase include gravita-tional potential energy and flow batteries, flywheel energy storage, energy storage in super capacitors, superconducting magnetic energy storage, and thermal energy storage. If these technologies are developed and deployed, they could also play a role in providing storage to enhance the capacity of existing and new variable electricity generation resources.

Flywheel energy accumulators are comprised of a massive or composite flywheel coupled with a motor generator and special brackets (often magnetic), set inside a housing at very low pressure to reduce self-discharge losses. They have a great cyc-ling capacity determined by fatigue design. From a practical point of view, electro-mechanical batteries are more useful for the production of energy in isolated areas. For example, some systems have been installed to supply areas of scattered houses as well as the islands of Scotland and Wales. In the first case, the batteries are used essentially to regulate and increase the quality of the current (constant and continuous voltage). Where supplying the islands is concerned, electromechanical batteries are used to ensure that a maximum of the energy consumed is generated by the local wind farms, and to improve the quality of supply when wind-turbine production is on the threshold of demand.

Energy storage in super capacitor components have both the characteristics of capacitors and electrochemical batteries, except that there is no chemical reaction, which greatly increases the cycling capacity. Energy storage in super capacitors is done in the form of an electric field between the two electrodes. This is the same principle as capacitors except that the insulating material is replaced by electrolyte ionic conductor in which ion movement is made along a conducting electrode with a very large specific surface. However, these storage technologies increase the already high capital cost of renewable energy sources, making them less economically attractive. It is a fact that market integration for renewable energy storage technologies has been slower than anticipated. The main reasons for this fact are the high capital cost for commercial storage technologies, space required for such infrastructures while the absence of a valuation framework for ancillary services furthermore discourages investments.

Superconducting magnetic energy storage is achieved by inducing DC current into a coil made of superconducting cables of nearly zero resistance, generally made of niobium titane (NbTi) filaments that operate at a very low temperature 270°C. The current increases when charging and decreases during discharge and has to be converted for AC or DC voltage applications. One advantage of this storage system is its great instantaneous efficiency, near 95 percent for a charge–discharge cycle. Moreover, these systems are capable of discharging the near totality of the stored energy, as opposed to batteries. They are very useful for applications requiring continuous operation with a great number of complete charge–discharge cycles. The fast response time (under 100 ms) of these systems makes them ideal for regulating network stability (load leveling). Their major shortcoming is the refrigeration system, which, while not a problem in itself is quite costly, makes operation more complicated (Mukherjee & Rao, 2019).

There are two types of thermal energy storage (TES) systems depending on whether they use sensible or latent heat. Latent-fusion-heat TES makes use of the liquid–solid transition of a material at constant temperature. During accumulation, the bulk material will shift from the solid state to liquid and, during retrieval, will transfer back to solid. The heat transfers between the thermal accumulator and the exterior environment are made through a heat transfer fluid. The energy is stored at a given temperature, the higher the heat the higher the concentration; the fusion enthalpy grows with the fusion temperature of the bulk material used. Setting up a sodium hydroxide, latent-heat accumulation systems in electric boilers could help limit demand for electrical power in industrial processes where the needs for steam are not continuous and vary in intensity. Sensible heat thermal storage is achieved by heating a bulk material (sodium, molten salt, pressurized water, etc.) that does not change states during the accumulation phase; the heat is then recovered to produce water vapor, which drives a turbo-alternator system. Using water as storage fluid involves high temperatures, above 200°C, making it impossible to store the water is a confined groundwater basin because irreparable damage to the ground would ensue. Very large volume watertight cisterns set in rock are needed. During off-peak hours, the hot water for storage can be obtained from a thermal plant, for example, condensation of the high-pressure steam from the boiler, or by tapping, at lower temperature, from the turbine outlets. Generating extra electricity during peak hours can be achieved by heating the water supply when retrieving stored energy and simultaneously reducing turbine outlet. A 5 percent overpower is obtained by an increase in steam output through the turbine (Cabeza, Castell, Barreneche, De Gracia, & Fernández, 2011). Drawing a parallel, cold storage is another type of thermal storage. It is used in case were renewable electricity surplus (or when the electricity is inexpensive) can be stored into industrial freezers lowering significantly the temperatures (e.g., –24°C), in terms of temperature, and when the cost of electricity is high let the temperature be slowly increased (e.g., up to –15°C) without actual electricity consumption (Xydis, 2013).

Lastly, a number of innovative applications act as energy storage solutions—although they are not defined by default as storage solutions. Electric vehicles (EVs) could play that role—as accumulated but distributed storage capacity of the network—as storing devices providing flexibility to the grid (Hu et al., 2014). Wind and solar electricity production volatility is partly "responsible" for the resulting curtailment and/or negative pricing in some grids (Fogelberg & Lazarczyk, 2017). Such type of storage can absorb the imbalances from high wind or solar integration and inject some of the locally stored electricity in EVs (based upon the vehicle-to-grid technology) to the grid together with the high amounts of wind. It should be mentioned that the consumers' acceptance of such solutions needs to be verified. Plant factories can be equally supportive to the grid. As in the EV case, if large-scale plant factories (e.g., large vertical farms) decide to participate under a demand response scheme, plant factories can adjust their consumption—or store in local batteries and consume later—based on signals they receive from the market. However, the last two solutions cannot be considered as usual (and traditional) storage solutions; however, it has been seen that recent trends introduce a number of nonconventional solutions (Xydis, Liaros, & Avgoustaki, 2020)

3.2.3 ENERGY STORAGE AND COSTS

The primary drawback to energy storage systems is the high capital cost. In the United States, there are several policy tools designed to help with providing cost competitiveness for battery installations. FERC [Federal Energy Regulatory Commission] Order 755, enacted in 2011, introduced a new method of evaluating prices for ancillary services and frequency regulation (Miller, Venkatesh, & Cheng, 2013; Nguyen, Byrne, Concepcion, & Gyuk, 2017). One of the effects of this legislation was financial incentive for technologies that could provide these services quickly, such as batteries and flywheels (Wesoff, 2013). FERC Order 784, enacted in 2013, expanded this pricing strategy's applicability to regulation services that are competitive in speed and accuracy. This legislation had the effect of "help[ing] utilities achieve rate recovery for energy storage equipment" (Wesoff, 2013).

Privately owned utility-scale batteries are eligible for some tax incentives. Specifically, battery owners can take advantage of a seven-year modified accelerated cost recovery system (MACR) for batteries with or without renewable energy charging the battery. If renewable energy is used to charge the battery between 75 and 99 percent of the time, the battery owner is eligible for a five-year MACRs as well as a portion of a 30 percent investment tax credit (ITC). If renewable energy is used to charge the battery all of the time, the battery owner is eligible for a five-year MACR as well as the entire 30 percent ITC (National Renewable Energy Lab, 2018).

High capital costs of storage can also be offset through direct government investment. Direct investment in research and development provides additional capability for research institutions, universities, and government organizations to work toward improving technologies and increasing manufacturing efficiency, with the end goal of reduced capital and operational and maintenance expenses for these new technologies. One example of direct government investment is the Department of Energy's (DOE) recent Energy Storage Grand Challenge, which provided a total of $187 million to national laboratories, corporations, research centers, and universities to work toward lowering cost and increasing the penetration of storage in the United States. Of the $187 million, $66 million were granted to battery-specific projects (Tsanova, 2020).

3.3   DISCUSSION AND CONCLUSIONS

The addition of energy storage infrastructure in the form of batteries and pumped storage hydropower has the potential to drastically expand the penetration of wind energy in the US market. Specifically, this strategy would allow curtailment of wind resources to be eliminated and wind power to provide all peaking power capacity. This strategy would allow wind energy to provide up to 26 percent of US electricity demand.

Energy storage infrastructure has the capability to smooth variability in both short (daily) and longer (seasonal) timescales. To examine the potential for storage to increase the penetration of wind energy in the United States, consider the energy lost due to curtailment of wind resources. On an annual basis, approximately 1–4 percent of wind generation is curtailed (Bird, Cochran, & Wang, 2014). In 2019, wind provided 7.3 percent of electricity generation or 300 billion kWh. Four percent of 300 billion kWh is equal to 12 billion kWh. Such a wasted electricity production could have provided energy to an average of 2.5–3 million US households (EWEA, 2021).

Energy storage infrastructure also has the capability to completely replace fossil fuel peaking power capacity. There is currently about 261 GW of peaking power capacity in the United States, which is primarily composed of simple cycle natural gas plants (Denholm, Nunemaker, Gagnon, & Cole, 2019). To determine an upper bound for increased penetration of wind energy, assume that all this capacity is replaced with wind plus storage installations, and then assume that those wind farms are run year-round at a 35 percent capacity factor. This provides an additional 800 billion kWh of electricity from wind annually.

In combination, eliminating curtailment and replacing fossil fuel peaking capacity with wind plus storage could contribute an additional 812 billion kWh of electricity generation from wind annually. If assuming the electricity demand in the United States remains constant, these 812 billion kWh represent an additional approximately 20 percent of annual electricity demand, raising the contribution from wind from 7 percent to a total of 27 percent of the national electricity generation.

Energy storage, in the form of lithium ion batteries, PSH, and newer technologies including hydrogen, CAES, gravitational potential energy storage, and flow batteries, has the potential to revolutionize the electricity sector and drastically increase penetration of variable electricity sources including wind. These technologies are well suited to smoothing both the short-term and long-term variability of wind power. The short-term variability of wind power is associated with weather forecasts and is therefore readily predictable. The long-term variability is harder to predict, but still follows general patterns from year to year. Energy storage is capable of smoothing both variations in production and can significantly increase the penetration of wind energy in the United States. If energy storage plus wind enables grid operators to eliminate curtailment, an additional 12 billion kWh of electricity from wind can be added to the grid each year. If energy storage plus wind is used to completely replace fossil fuel peaking capacity, and additional 800 billion kWh of electricity from wind can be added to the grid each year. In sum, these contributions add to 812 billion kWh annually, or 20 percent of the United States' 2019 electricity generation.

Investment at industrial, research, and governmental levels is required to decrease the cost of storage technologies, expedite the environmental impact assessment process, and remove other barriers to the construction of energy storage infrastructure. Furthermore, technology to recycle lithium on a broad scale is imperative to the widespread adoption and continued use of lithium ion batteries for storage and other applications such as electric vehicles.

The potential contribution from batteries is large, but investment and research efforts are crucial to driving these technologies forward to large-scale market penetration. Large-scale market penetration of energy storage plus wind installations could also curry public favor of wind to a larger degree than is experienced currently. Future legislation surrounding energy storage and renewable energy will heavily influence the rate of adoption. Legislation that promotes economic competitiveness and direct financial support for research and development and supply chain development is crucial to support continued growth of energy storage installations. The shape and speed of adoption of this technology will be influenced by legislation, overall energy transition progress, project development, fuel prices, and numerous other factors. However, increased energy storage installation is imperative to provide flexibility from variable electricity generation, improving penetration of renewable energy, and moving the United States closer to a 100 percent renewable electricity generation profile.

REFERENCES

Andrews, J., & Jelley, N. (2017). *Energy Science: Principles, Technologies, and Impacts.* Oxford: Oxford University Press.

Apostolou, D., Casero, P., Gil-Hernández, V., & Xydis, G. (2021). Integration of a light mobility urban scale hydrogen refuelling station for cycling purposes in the transportation market. *International Journal of Hydrogen Energy*, 46(7), 5756–5762.

Bird, L., Cochran, J., & Wang, X. (2014). *Wind and Solar Energy Curtailment: Experience and Practices in the United States*. Golden: National Renewable Energy Laboratory.

Cabeza, L. F., Castell, A., Barreneche, C. D., De Gracia, A., & Fernández, A. I. (2011). Materials used as PCM in thermal energy storage in buildings: A review. *Renewable and Sustainable Energy Reviews*, 15(3), 1675–1695.

Denholm, P., Nunemaker, J., Gagnon, P., & Cole, W. (2019). *The Potential for Battery Energy Storage to Provide Peaking Capacity in the United States*. Golden: National Renewable Energy Laboratory.

Enevoldsen, P., Permien, F.-H., Bakhtaoui, I. v.-K., Jacobson, M. Z., Xydis, G., Sovacool, B. K.,… Oxley, G. (2019). How much wind power potential does Europe have? Examining European wind power potential with an enhanced socio-technical atlas. *Energy Policy*, 132, 1092–1100.

EWEA. (2021). *Wind Energy Basics. Wind Energy's Frequently Asked Questions*. Retrieved from EWEA: www.ewea.org/wind-energy-basics/faq/

Fogelberg, S., & Lazarczyk, E. (2017). Wind power volatility and its impact on production failures in the Nordic electricity market. *Renewable Energy*, 105, 96–105.

Gonzales-Salazar, M. A., Kirsten, T., & Prchlik, L. (2018). Review of the operational flexibility and emissions of gas- and coal-fired power plants in a future with growing renewables. *Renewable and Sustainable Energy Reviews*, 82, 1497–1513.

Harper, G., Sommerville, R., Kendrick, E., Driscoll, L., Slater, P., Stolkin, R., … Anderson, P. (2019). Recycling lithium-ion batteries from electric vehicles. *Nature*, 575, 76–86.

Hu, J., Morais, H., Zong, Y., You, S., Bindner, H. W., Wang, L., & Wu, Q. (2014). Multi-agents based modelling for distribution network operation with electric vehicle integration. *Intelligent Computing in Smart Grid and Electrical Vehicles*, 136, 349–358.

International Energy Agency. (2019, May). *Tracking Energy Integration: Energy Storage*. Retrieved from IEA: www.iea.org/reports/tracking-energy-integration/energy-storage

IRENA. (2017, October). *Electricity Storage and Renewables: Costs and Markets to 2030*. Retrieved from IRENA. www.irena.org/DocumentDownloads/Publications/IRENA_Electricity_Storage_Costs_2017.pdf

IRENA. (2019). *Innovation Landscape Brief: Utility-Scale Batteries*. Retrieved from International Renewable Energy Agency: www.irena.org/-/media/Files/IRENA/Agency/Publication/2019/Sep/IRENA_Utility-scale-batteries_2019.pdf

Jacoby, M. (2019, July 14). *It's Time to Get Serious about Recycling Lithium-Ion Batteries*. Retrieved from *Chemical & Engineering News*. https://cen.acs.org/materials/energy-storage/time-serious-recycling-lithium/97/i28

Karabiber, O., & Xydis, G. (2020). Forecasting day-ahead natural gas demand in Denmark. *Journal of Natural Gas Science & Engineering*, 76, 103193.

Laporte, A. (2019, February 22). *Fact Sheet: Energy Storage* (2019). Retrieved from Environmental and Energy Study Institute. www.eesi.org/papers/view/energy-storage-2019

Mayes, F. (2019, October 31). *Most Pumped Storage Electricity Generators in the U.S. Were Built in the 1970s*. Retrieved from U.S. EIA: www.eia.gov/todayinenergy/detail.php?id=41833

Menéndez, J., & Fernández-Oro, J. M. (2020). Efficiency analysis of underground pumped storage hydropower plants. *Journal of Energy Storage*, 28, 101234.

Merchant, E. F. (2019, March 9). *US Wind Industry Frets as Major Transmission Lines Stall*. Retrieved from GreenTech Media: www.greentechmedia.com/articles/read/an-argument-as-old-as-wind-the-transmission-conundrum#gs.8et6tv

Michalitsakos, P., Mihet-Popa, L., & Xydis, G. (2017). A hybrid RES distributed generation system for autonomous islands: A DER-CAM and storage-based economic and optimal dispatch analysis. *Sustainability*, 9(11), 2010.

Miller, R., Venkatesh, B., & Cheng, D. (2013). Overview of FERC Order No. 755 and proposed MISO implementation. *IEEE Power & Energy Society General Meeting*, 1–5.

Mongird, K., Viswanathan, V., Balducci, P., Alam, J., Fotedar, V., Koritarov, V., & Hadjerioua, B. (2019, July). *Energy Storage Technology and Cost Characterization Report.* Retrieved from Energy.gov: www.energy.gov/sites/prod/files/2019/07/f65/Storage%20 Cost%20and%20Performance%20Characterization%20Report_Final.pdf

Mukherjee, P., & Rao, V. V. (2019). Design and development of high temperature superconducting magnetic energy storage for power applications—A review. *Physica C: Superconductivity and Its Applications*, 563, 67–73.

National Renewable Energy Lab. (2018, January). *Federal Tax Incentives For Energy Storage Systems.* Retrieved from NREL. www.nrel.gov/docs/fy18osti/70384.pdf

Nguyen, T. A., Byrne, R. H., Concepcion, R. J., & Gyuk, I. (2017). Maximizing revenue from electrical energy storage in MISO energy & frequency regulation markets. *IEEE Power & Energy Society General Meeting*, 1–5.

Panagiotidis, P., Effraimis, A., & Xydis, G. (2019). An R-focused forecasting approach for efficient demand response strategies in autonomous micro grids, energy & environment. *Energy & Environment*, 30(1), 63–80.

Peake, S. (2018). *Renewable Energy: Power for a Sustainable Future.* Oxford: Oxford University Press.

Soret, A., Torralba, V., Cortesi, N., Christel, I., Palma, L., Manrique-Sunen, A., ... Doblas-Reyes, F. (2019). Sub-seasonal to seasonal climate predictions for wind energy forecasting. *Journal of Physics: Conference Series*, 1222, 1–6.

Strbac, G. (2008). Demand side management: Benefits and challenges. *Energy Policy*, 36, 4419–4426.

Tsanova, T. (2020, February 12). *Battery Storage R&D Projects Get USD 66m under US DOE Programme.* Retrieved from Renewables Now. https://renewablesnow.com/news/ battery-storage-rd-projects-get-usd-66m-under-us-doe-programme-686988/

U.S. Department of Energy. (2019). *2018 Wind Technologies Market Report.* Oak Ridge: U.S. Department of Energy.

U.S. Department of Energy. (n.d.). *U.S. Installed and Potential Wind Power Capacity and Generation.* Retrieved from Energy.gov Office of Energy Efficiency & Renewable Energy: https://windexchnage.energy.gov/maps-data/321

U.S. EIA. (2020, February 21). *Hourly Electricity Consumption Varies Throughout the Day and Across Seasons.* Retrieved from U.S. EIA: www.eia.gov/todayinenergy/ detail.php?id=42915

U.S. EIA. (2020, February 7). *What Is U.S. Electricity Generation by Energy Source?* Retrieved from U.S. EIA: www.eia.gov/tools/faqs/faq.php?id=427&t=3

U.S. Energy Information Administration. (2019a). *Levelized Cost and Levelized Avoided Cost of New Generation Resources in the Annual Energy Outlook 2019.* Washington, DC: Independent Statistics and Analytics.

U.S. Energy Information Administration. (2019b, April 4). *Wind Explained: Where Wind Power Is Harnessed.* Retrieved from U.S. Energy Information Administration: www.eia.gov/ energyexplained/wind/where-wind-power-is-harnessed.php

Ucal, S. M., & Xydis, G. (2020). Multidirectional relationship between energy resources, climate changes and sustainable development: Technoeconomic analysis. *Sustainable Cities and Society*, 60, 102210.

Ulazia, A., Saenz, J., Ibarra-Berastegi, G., Gonzales-Roji, S. J., & Carreno-Madiabeitia, S. (2019). Global estimations of wind energy potential considering seasonal air density changes. *Energy*, 187, 115938.

USGS. (2020, January 31). *Mineral Commodity Summaries 2020*. Retrieved from U.S. Department of the Interior, U.S. Geological Survey: https://pubs.usgs.gov/periodicals/mcs2020/mcs2020.pdf

Wan, Y. H. (2012). *Long-Term Wind Power Variability*. Golden: National Renewable Energy Laboratory.

Wang, J., You, S., Zong, Y., Cai, H., Træholt, C., & Dong, Z. Y. (2019). Investigation of real-time flexibility of combined heat and power plants in district heating applications. *Applied Energy*, 237, 196–209.

Wesoff, E. (2013, August 12). *FERC's Energy Storage Ruling Could Jump-Start Big Batteries*. Retrieved from Greentech Media: www.greentechmedia.com/articles/read/FERCs-Energy-Storage-Ruling-Could-Jump-Start-Big-Batteries

Xydis, G. (2012). Effects of air psychrometrics on the exergetic efficiency of a wind farm at a coastal mountainous site – An experimental study. *Energy*, 37, 632–638.

Xydis, G. (2013). Wind energy to thermal and cold storage—A systems approach. *Energy and Buildings*, 56, 41–47.

Xydis, G., Liaros, S., & Avgoustaki, D. (2020). Small scale plant factories with artificial lighting and wind energy microgeneration: A multiple revenue stream approach. *Journal of Cleaner Production*, 120227.

Xydis, G., Pechlivanoglou, G., & Nayeri, N. C. (2015). Wind turbine waste heat recovery—A short-term heat loss forecasting approach. *Challenges*, 6, 188–201.

Zhang, G., Li, Y., Daemen, J. J., Yang, C., Wu, Y., Zhang, K., & Chen, Y. (2015). Geotechnical feasibility analysis of compressed air energy storage (CAES) in bedded salt formations: A case study in Huai'an City, China. *Rock Mechanics and Rock Engineering*, 48(5), 2111–2127.

4 Investigating the Electricity Portfolio Optimization for Renewable Energy Sources

Fazıl Gökgöz and Ahmet Yıldırım Erdoğan

4.1 INTRODUCTION

Electricity, which is a product necessary for the continuation of human daily life, also plays an important role in the growth of industry, trade, and countries themselves.

While some energy sources are renewable—solar, wind, hydraulic, and geothermal energies—others are non-renewable energy sources, including coal, petroleum, and natural gas, also known as fossil fuels. When the non-renewable and renewable resources used in electricity production are examined, it is clear that the importance

DOI: 10.1201/9781003240129-4

of renewable energy resources is increasing. The fact that these resources are renewable, and that there are fewer negative environmental factors due to their use, causes an increase in demand. Even though renewable energies are more expensive and more limited than fossil fuels, they are the least harmful types of energy; the use of non-renewable energy leads to harmful wastes forming and harmful gas emissions being released into the environment (Bekar, 2020).

Electricity appears as a commodity that has different markets and that can be bought and sold, differing from other commodities since it has unique features such as instantaneous generation, consumption obligation, and the inability to store.

Although non-renewable energy sources are still available in nature, they are limited because they will run out over time (Bekar, 2020). The generation costs of electricity produced by renewable energy sources are more expensive than the generation cost of electricity produced with fossil sources, except hydraulic resources (Ağır et al., 2020; BOUN, 2021). Therefore, renewable energies are more limited than fossil fuels (Bekar, 2020). In Turkey, the biggest obstacle for the expansion of renewable energy is the financial and technological constraints. An incentive system to increase domestic production has not yet been established. In addition, difficulties arising from bureaucratic procedures are the biggest obstacle for investors who want to invest in this field. Finding skilled workers in the sector is also an important problem. Especially entrepreneurs who want to produce without a license and meet their own needs give up because of these obstacles (Kayışoğlu & Diken, 2019). Therefore, it is necessary to use the limited non-renewable and renewable resources effectively and efficiently. Under these constraints, suppliers in the electricity market have to deal with the electricity market's risks, including the technical, transmission, human resources, and operational risks of electricity generation. In order for the producers to manage their risks more effectively in the free electricity market environment, they need to optimize their capacity to generate and market bid strategies and to determine the strategies for production, investment, and marketing in the short, medium, and long term, taking into account market conditions and risks.

In an environment of intense competition, the main purpose of electricity generation companies is to maximize profits while minimizing market rivals. Electricity suppliers need to establish a trading strategy with risk management before bidding on the electricity market, whether it is a spot market or a bilateral contract market. The main target at this point is to determine portfolio weights. For this determination, it is necessary to optimize the expected risks and expected returns.

Portfolio management has been developing in finance theory in the past, with new approaches developed in the field of securities management every day. According to many scientists, Harry M. Markowitz brought a new dimension to portfolio management approaches with his 1952 article "Portfolio Selection," laying the foundation of modern portfolio theory (MPT) (Gökgöz & Atmaca, 2012; 2017). In the classical portfolio approach, it was argued that an unsystematic risk could be reduced by investing in stocks, gold, foreign currency, bonds, and bills in different sectors, as well as creating a portfolio basket with a large number of securities; the correlation between securities was not considered (Statman, 1987; Copeland et al., 2005; Gökgöz & Atmaca, 2012). In MPT, it has been stated that, in order to reduce the nonsystematic risk, the correlation between entities should be taken into consideration along with diversification (Markowitz, 1952; Liu & Wu, 2006; Gökgöz & Atmaca, 2012). In this context, Markowitz handled the concepts of risk and return together, succeeding in making the risk measurable.

While mean variance was used as the measure of risk in the mean variance optimization (MV) specified in the MPT, Konno and Yamazaki used the mean absolute deviation as the risk measure, introducing the concept of mean absolute deviation optimization (MAD) to the literature (Konno & Yamazaki, 1991). Thus, the problem of portfolio selection has moved from a quadratic program to a linear program (Simaan, 1997). The quadratic programming method proposed in MPT—a classical approach for the portfolio selection problem—is based on normal distribution and the assumptions that the investor is risk-averse due to the operation difficulties, the need for much computation time, and the MAD model as an alternative to this model (Kardiyen, 2007). These two models have been used extensively in financial markets and the pairwise comparison results have been shared for the benefit of decision makers (Kardiyen, 2007; Kasenbacher et al., 2017).

Turkey is located on an important geothermal area and has over 1,300 geothermal resources. In this regard, Turkey is the leading country in Europe and the seventh country in the world in terms of geothermal resources (Takan & Kandemir, 2020). In addition, the importance of renewable energy is increasing daily compared to non-renewable energy. For these reasons, in this empirical study, for geothermal power plants in the Turkish electricity market, the MV and MAD optimization models have been adapted. Each model was subjected to a total of 12 optimizations using six different objective functions: minimum risk portfolio, maximum utility ($A = 3$, $A = 4$, $A = 5$) portfolio, maximum Sharpe ratio portfolio, and maximum return portfolio. The Sharpe ratio is used to calculate the performance of portfolios determined as a result of optimization. The results found were compared for both models, and suggestions were made to the decision makers of the electricity market.

4.2 LITERATURE

With modern portfolio theory, scientists have increased their work on portfolio optimization. As a result of their studies, they have revealed different optimization models and theories. Some of the financial optimization models, which are frequently used in the finance sector and have found a wide application area, have now become applicable to different sectors. Financial optimization models started to be used to model and solve the sales portfolios of electricity suppliers operating in the electricity market, which is one of the nonfinancial sectors.

When the issue of applying financial optimization models in electricity markets is examined in the literature, it is seen that the studies have intensified after the year 2000.

Byström (2003) conducted a study on minimizing the variance in the Nord Pool electricity market and spot and future market prices were used as the data set. The study discusses electricity futures and how they can be used to hedge short-term risks in spot market positions, as well as the minimum variance protection rate and how it can be calculated in various ways.

Dahlgren et al. (2003) made optimizations to the SP15 California PX electricity market with VaR/CVaR models (95 percent and 99 percent, respectively), conducting a relatively simple and primitive study for the electricity market.

Liu (2004) applied portfolio optimization between the mean-variance model in the California electricity market and the spot market—hedged with futures and forward applications—and an application that takes fuel prices into account emerged.

Liu and Wu (2006) conducted three separate case studies in the California electricity market with the mean-variance model, taking into account the spot market, fuel prices, and contract prices. Liu and Wu (2007a) applied portfolio optimization between three regional and one spot market assets using the mean-variance model in the Pennsylvania, New Jersey, and Maryland (PJM) electricity market. Liu and Wu (2007b), again using the PJM electricity market data, analyzed the performance of 10 regions' bilateral agreements (risk-free) with the vector auto-regression (VaR) model and two different business portfolio situations within the scope of a spot market asset approach, which is risky.

Feng et al. (2007) carried out portfolio optimization in the PJM electricity market using the mean-variance optimization model over the day-ahead market, risky bilateral agreement, and risk-free bilateral agreement.

In Munoz et al. (2009), using Spanish electricity market data, the mean-variance optimization model determined the renewable energy resources that should be included in the optimum investment portfolio for Spain, according to nine different situations.

Pindoriya et al. (2010) explain that, in the PJM electricity market, using the mean-variance-skewness model and PJM electricity market data, the electricity generation cost function was defined quadratically, and asset placement optimization was applied between the two contracts and spot markets.

Kazempour and Moghaddam (2011) optimized using the Spanish electricity market data with variance, covariance, and conditional value at risk (CVaR) models. In the study, the weekly optimization of a power plant under the constraints of fuel price, fuel storage area, consumption amount, emission amount, electricity market, and additional production capacity market was produced, as was the expected return effective limit based on certain standard deviation and risk penalty factors.

Bhattacharya and Kojima (2012) applied the mean-variance optimization model to the Japanese electricity market in 2012. As a result of the application, it is shown that the renewable energy portfolio for Japan can reach up to 9 percent.

Suksonghong et al. (2014) conducted the analysis of the energy sales portfolios for nine different regions over two sample cases using the PJM electricity market data as a multiobjective optimization with the mean-variance-skewness model.

Boroumand et al. (2015) applied VaR and CVaR optimization models on the French electricity market in his study. In this study, considering both market price uncertainty and customer consumption uncertainties, portfolios that minimize risks were sought from the perspective of electricity suppliers. Bilateral agreements, production, and market shares have been researched.

Gökgöz and Atmaca (2012) applied mean-variance optimization model to the Turkish electricity market. Within the scope of the study, they carried out portfolio optimization studies with different Turkish electricity assets in various case studies.

Gökgöz and Atmaca (2017) investigated the mean-variance, the down side, and semi-variance optimization models for the Turkish electricity markets in different empirical case studies.

In this empirical study, the MV and MAD optimization models were applied to the Turkish electricity market. Electricity generation cost was used for geothermal power plants within the scope of the study. It has been considered as a 24-hour risk asset, with six different objective functions for both models—minimum risk portfolio, maximum utility ($A = 3$, $A = 4$, $A = 5$) portfolio, maximum Sharpe ratio portfolio, and maximum return portfolio. A total of 12 optimization studies were carried out; the MV and MAD optimization models were compared in detail, and suggestions were made to the electricity market decision makers for both models.

4.2.1 MPT AND MV OPTIMIZATION MODEL

The classical portfolio theory notes that the risk can be minimized by diversification. In this approach, diversification is done in an unsystematic manner, regardless of the correlation between the assets that compose the portfolio (Copeland et al., 2005; Gökgöz & Atmaca, 2012). Harry M. Markowitz brought a new dimension to portfolio management approaches with his 1952 article "Portfolio Selection," in which he laid the foundations for the modern portfolio theory (MPT). MPT states that, in order to reduce the nonsystematic risk, the correlation between entities should be taken into consideration along with diversification (Markowitz, 1952; Liu & Wu, 2006; Gökgöz & Atmaca, 2012). MPT relies on the optimization of MV, which provides minimum risk for an expected level of return or maximum return for an expected value of risk, while also seeking effective portfolios (Gökgöz, 2009; Gökgöz & Atmaca, 2017). The basic assumptions of MPT are listed below (Defusco et al., 2004; Huang & Wu, 2016; Gökgöz & Atmaca, 2017):

- Investors have all the information about expected returns, variances, and covariance of risky assets.
- Investors only consider the expected returns, variances, and covariances of risky assets.
- All investors avoid risk.
- The expected returns of assets are normally distributed.
- There are not any taxes or transaction costs.

According to MPT, the return distribution of portfolio alternatives can only be created by using average returns and variances. In this context, the expected returns, variances, and covariance matrix of assets should be created in order to perform MV optimization (Gökgöz & Atmaca, 2012, 2017).

Markowitz called the portfolios that provide maximum return at the desired risk level or have minimum risk at the desired level of return "the effective portfolio," and the curve that combines effective portfolios in the risk–return graph "the efficient frontier" (Copeland et al., 2005; Kardiyen, 2007).

As seen in Figure 4.1, the minimum risk portfolio is the lowest risk portfolio that can be created with existing assets. The section above the minimum risk portfolio point is the region where effective results are sought for the solution. Therefore,

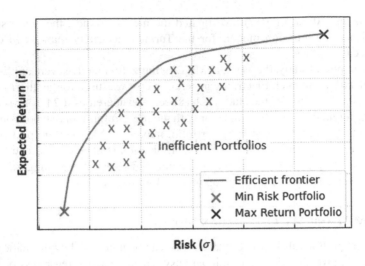

FIGURE 4.1   The efficient frontier of the MV model.

the efficient frontier curve is started with the minimum risk portfolio point. While searching for a solution portfolio, portfolios on the efficient frontier starting with this point are taken into consideration.

In MV optimization, the expected return, variance, and covariance matrix of the portfolio to be created is calculated as follows (Huang & Wu, 2016; deLlano-Paz et al., 2017; Wattoo et al., 2020):

$$E\left(R_p\right)=\sum_{i=1}^{n}W_iR_i \tag{4.1}$$

$$\sigma_p^2=\sum_{i=1}^{n}\sum_{j=1}^{n}W_iW_j\sigma_{ij} \tag{4.2}$$

$$\text{Covariance Matrix}=\begin{bmatrix} \sigma_1^2 & \sigma_{1,2} & \cdots & \sigma_{1,n} \\ \sigma_{2,1} & \sigma_2^2 & & \\ \vdots & & \ddots & \\ \sigma_{n,1} & & & \sigma_n^2 \end{bmatrix} \tag{4.3}$$

where n is the number of current risky assets, W_i is the asset i's weight in the portfolio, W_j is the asset j's weight in the portfolio, R_i is the asset i's return, σ_{ij} is the value of covariance between assets i and j, $E(R_p)$ is the portfolio's expected return, σ_p^2 is the portfolio's variance.

In MV optimization, the minimum risk portfolio with the minimum risk objective function is calculated as follows (Gökgöz & Atmaca, 2012; 2017; Huang & Wu, 2016):

$$Min.\left(\sigma_p^2\right) = \sum_{i=1}^{n} \sum_{j=1}^{n} W_i W_j \sigma_{ij} \tag{4.4}$$

s.t.

$$\sum_{i=1}^{n} W_i R_i = R_{\text{target}} \tag{4.5}$$

$$\sum_{i=1}^{n} W_i = 1 \tag{4.6}$$

$$0 \leq W_i \leq 1, \left[i = 1, 2, ..., n\right] \tag{4.7}$$

In MV optimization, the maximum return portfolio with the maximum return objective function is calculated as follows (Gökgöz, 2009; Ivanova & Dospatliev, 2017):

$$Max.E\left(R_p\right) = \sum_{i=1}^{n} W_i R_i \tag{4.8}$$

s.t.

$$\sum_{i=1}^{n} W_i = 1 \tag{4.9}$$

$$0 \leq W_i \leq 1, \left[i = 1, 2, ..., n\right] \tag{4.10}$$

In MV optimization, the maximum Sharpe ratio portfolio is calculated as follows, with the maximum Sharpe ratio objective function (Liu & Wu, 2007a; Zhu et al., 2011; Cucchiella et al., 2016; deLlano-Paz et al., 2017):

$$Max.\left(\text{Sharpe Ratio}\right) = \frac{E\left(R_p\right) - r_f}{\sigma_p^2} \tag{4.11}$$

s.t.

$$\sum_{i=1}^{n} W_i = 1 \tag{4.12}$$

$$0 \leq W_i \leq 1, \left[i = 1, 2, \ldots, n \right] \tag{4.13}$$

where r_f represents the rate of the risk-free asset's return.

In MV optimization, the maximum utility portfolio with the maximum utility objective function is calculated as follows (Liu, 2004; Liu & Wu, 2006; Liu & Wu, 2007a; Gökgöz & Atmaca, 2012; Huang & Wu, 2016; deLlano-Paz et al., 2017; Wattoo et al., 2020):

$$\text{Max.}(U) = E\left(R_p \right) - \frac{1}{2} A \sigma_p^2 \tag{4.14}$$

s.t.

$$\sum_{i=1}^{n} W_i = 1 \tag{4.15}$$

$$0 \leq W_i \leq 1, \left[i = 1, 2, \ldots, n \right] \tag{4.16}$$

where U represents the utility function. In the utility function, the coefficient A represents the risk aversion degree of the investor. A small coefficient of A represents that the investor likes risk, while a large coefficient represents risk aversion. In general, the value of the coefficient A is 3 to represent the average level of risk aversion in financial studies (Gökgöz & Atmaca, 2012). It is understood that the investor likes the higher-than-normal risk when $A < 3$, and the investor avoids the higher-than-normal risk when $A > 3$ (Liu & Wu, 2007a; Gökgöz & Atmaca, 2012).

4.3   MAD OPTIMIZATION MODEL

The mean absolute deviation (MAD) portfolio optimization model was proposed by Konno and Yamazaki in the 1991 MPT report as an alternative to the MV portfolio optimization model defined by Markowitz. As a measure of risk, the MAD model uses the mean absolute deviation. Therefore, Konno and Yamazaki used mean absolute deviation in the MAD model instead of the variance considered to be minimized in Markowitz's MV model. Thus, portfolio optimization has transformed from quadratic programming to linear programming, and there have been attempts to overcome the computational difficulties in linear programming by using quadratic programming in large portfolios (Konno & Yamazaki, 1991; Simaan, 1997; Kardiyen, 2007; 2008).

The MAD optimization model's advantages can be listed as follows:

- In the MAD model, the returns do not have to show a normal distribution, as it does not require any distribution assumptions (Mansini et al., 2003).
- With the MAD model, portfolio optimization is done by linear programming. Therefore, the portfolio selection problem is easier to solve (Konno & Yamazaki, 1991).

- To establish the MAD model, a variance–covariance matrix doesn't need to be developed. When new assets are added, the model can be easily updated (Kardiyen, 2007; 2008).
- MAD model cannot contain more than $2T + 2$ assets in the optimal portfolio solution. Therefore, if there is a need to limit the number of assets in the portfolio, T can be used as a control variable (Kardiyen, 2007, 2008).
- In the MAD model, a solution is always produced, even if the returns of all assets are negative within the same period (Kardiyen, 2007, 2008).
- The MAD model's optimal solution at the efficient frontier has the feature of making the expected benefit the highest independent of the distribution of returns (Kardiyen, 2007, 2008).

While not needing to create a variance–covariance matrix in the MAD model provides ease of operation, there are also opinions that it leads to great estimation risks (Kardiyen, 2007, 2008).

As a measure of the expected return and risk of the portfolio to be generated with the MAD portfolio selection model, the mean absolute deviation value is calculated as follows (Kardiyen, 2008; 2007; Konno & Koshizuka, 2005; Konno & Yamazaki, 1991):

$$E(R(x)) = \sum_{j=1}^{n} R_j x_j \tag{4.17}$$

$$w(x) = E\left[\left|\sum_{j=1}^{n} R_j x_j - E\left[\sum_{j=1}^{n} R_j x_j\right]\right|\right] \tag{4.18}$$

where $E(R(x))$ represents the portfolio's expected return, R_j represents the asset j's return, x_j represents the asset j's weight in the portfolio, and $w(x)$ represents the average absolute deviation value.

In MAD optimization, the minimum risk portfolio with the minimum risk objective function is calculated as follows (Kardiyen, 2008; 2007; Konno & Koshizuka, 2005; Konno & Yamazaki, 1991):

$$\text{Min.} w(x) = E\left[\left|\sum_{j=1}^{n} R_j x_j - E\left[\sum_{j=1}^{n} R_j x_j\right]\right|\right] \tag{4.19}$$

s.t.

$$\sum_{j=1}^{n} E(R_j) x_j \geq R_{\text{target}} \tag{4.20}$$

$$\sum_{j=1}^{n} x_j = 1 \tag{4.21}$$

$$0 \le x_j \le 1, \left[j = 1,2,\ldots,n \right] \tag{4.22}$$

where R_{target} represents the minimum rate of return the investor is targeting and $E(R_j)$ represents the expected rate of return on asset j.

r_{jt}; time period t is the yield obtained for ($t = 1,2,\ldots, T$). It is assumed that r_{jt} can be derived from historical data or some future predictions. In addition, the expected value of the random variable is supposed to converge with the average of the data obtained.

$$r_j = E(R_j) = \sum_{t=1}^{T} r_{jt} / T \tag{4.23}$$

In this case, $w(x)$ is converged as follows.

$$w(x) = E \left[\left| \sum_{j=1}^{n} R_j x_j - E \left[\sum_{j=1}^{n} R_j x_j \right] \right| \right] = \frac{1}{T} \sum_{t=1}^{T} \left| \sum_{j=1}^{n} \left(r_{jt} - r_j \right) x_j \right| \tag{4.24}$$

where $a_{jt} = r_{jt} - r_j$; $j = 1, 2, \ldots, n$; For $t = 1,2,\ldots, T$, the minimum risk objective function is calculated as follows:

$$\text{Min.} w(x) = \sum_{t=1}^{T} \left| \sum_{j=1}^{n} a_{jt} x_j \right| / T \tag{4.25}$$

s.t.

$$\sum_{j=1}^{n} r_j x_j \ge R_{target} \tag{4.26}$$

$$\sum_{j=1}^{n} x_j = 1 \tag{4.27}$$

$$0 \le x_j \le 1, \left[j = 1,2,\ldots,n \right] \tag{4.28}$$

The minimum risk objective function has become linear as a result of the definitions. The model above is equivalent to the linear programming model that follows (Konno & Yamazaki, 1991):

$$\text{Min.}w(x) = \sum_{t=1}^{T} y_t / T \qquad (4.29)$$

s.t.

$$y_t + \sum_{j=1}^{n} a_{jt} x_j \geq 0 \,, t = 1,2,\ldots,T \qquad (4.30)$$

$$y_t - \sum_{j=1}^{n} a_{jt} x_j \geq 0 \,, t = 1,2,\ldots,T \qquad (4.31)$$

$$\sum_{j=1}^{n} r_j x_j \geq R_{\text{target}} \qquad (4.32)$$

$$\sum_{j=1}^{n} x_j = 1 \qquad (4.33)$$

$$0 \leq x_j \leq 1, \; [j = 1,2,\ldots,n] \qquad (4.34)$$

In MAD optimization, the maximum return portfolio with the maximum return objective function is calculated as follows:

$$\text{Max.}E(R(x)) = \sum_{j=1}^{n} R_j x_j \qquad (4.35)$$

s.t.

$$\sum_{j=1}^{n} x_j = 1 \qquad (4.36)$$

$$0 \leq x_j \leq 1, \; [j = 1,2,\ldots,n] \qquad (4.37)$$

In MAD optimization, the maximum Sharpe ratio portfolio is calculated as follows, with the maximum Sharpe ratio objective function:

$$\text{Max.}(\text{Sharpe Ratio}) = \frac{E(R(x)) - r_f}{w(x)} \qquad (4.38)$$

s.t.

$$\sum_{j=1}^{n} x_j = 1 \qquad (4.39)$$

$$0 \leq x_j \leq 1, \left[j = 1, 2, \ldots, n \right] \qquad (4.40)$$

where r_f represents the rate of the risk-free asset's return.

In MAD optimization, the maximum utility portfolio with the maximum utility objective function is calculated as follows:

$$\text{Max.}(U) = E(R(x)) - \frac{1}{2} w(x) \qquad (4.41)$$

s.t.

$$\sum_{j=1}^{n} x_j = 1 \qquad (4.42)$$

$$0 \leq x_j \leq 1, \left[j = 1, 2, \ldots, n \right] \qquad (4.43)$$

where U represents the utility function. In the utility function, the coefficient A represents the risk aversion degree of the investor.

4.4 RENEWABLE ENERGY RESOURCES OF TURKEY

In fact, 86 percent of the primary energy resources consumed in the world are provided from fossil-based fuels (coal, petroleum, and natural gas). The formation of these fuels occurs over millions of years; because of these features, they are called non-renewable energy sources. In addition, fossil fuels are expensive and have a negative impact on the environment. This is especially true for developing countries that import fossil-based fuels, spending a large part of their budgets on the import bill. For these reasons, the demand for primary energy sources that can be alternatives to fossil-based fuels has increased in recent years. Renewable energies such as hydraulics, solar, wind, geothermal, and biomass are the leading sources (Kayışoğlu & Diken, 2019).

Renewable energy sources are more preferable to fossil fuels because they are environmentally friendly, reliable, and have unlimited reserves. Population growth and technological developments increase energy consumption day by day, so the use of renewable energy resources is increasing due to the fact that non-renewable energy resources will be depleted in the near future, causing environmental problems and external dependence (Takan & Kandemir, 2020).

After special assessment, Turkey's total electricity generation was determined to be 252.172 GWh at end of October 2020. Of this, 113.25 GWh of total electricity generation was provided from renewable energy sources, and 138.922 GWh came from non-renewable energy sources. The ratio of energy generated by using renewable resources to total energy generation is 44.91 percent (Energy Portal, 2021). Figure 4.2 shows the total energy production rates according to energy sources.

As seen in Figure 4.2, as of the end of October 2020, 70.836 GWh (28.09 percent) was generated using hydraulic resources, 19.992 GWh (7.93 percent) using wind resources, 10.380 GWh (4.12 percent) using solar resources, 7.626 GWh (3.03 percent) using geothermal resources, and 4.418 GWh (1.75 percent) using waste and garbage resources. The ratio of energy produced by using non-renewable resources to total energy production is 55.09 percent (Energy Portal, 2021).

Turkey's total electricity installed capacity is 93.282 MWe as of the end of October 2020 and 50.36 percent of the total installed electrical energy power belongs to renewable energy resources. The total installed power of energy plants where renewable resources are used is 46.976 MWe (Energy Portal, 2021).

As seen in Figure 4.3, as of the end of October 2020, the installed power of hydraulic power plants is 29,632 MWe (32.02 percent), the installed power of wind power plants is 8,077 MWe (8.61 percent), the installed power of solar power plants is 6,408 MWe (6.79 percent), the installed power of geothermal power plants is 1,515 MWe (1.63 percent), and the installed power of waste and garbage power plants

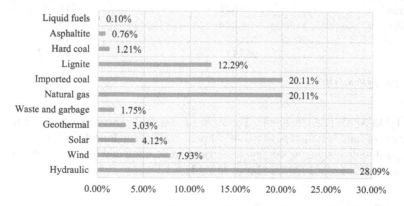

FIGURE 4.2 Turkey's electricity generation rates chart by resources at the end of October 2020 (Energy Portal, 2021).

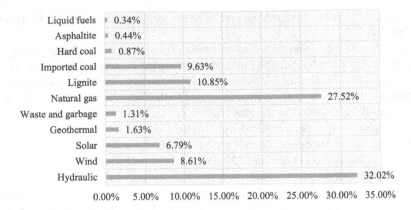

FIGURE 4.3 Turkey's electricity installed power rates chart by resources at the end of October 2020 (Energy Portal, 2021).

TABLE 4.1
Turkey's Electricity Energy Generation and Installed Power Chart by Resources at the End of October 2020

Resources	Generation at the end of October 2020 (GWh)	Installed power at the end of October 2020 (MWe)
Hydraulic	70.836	29.632
Wind	19.992	8.077
Solar	10.380	6.408
Geothermal	7.626	1.515
Waste and garbage	4.418	1.238
Natural gas	52.023	25.632
Lignite	30.995	10.097
Imported coal	50.705	8.967
Hard coal	3.047	811
Asphaltite	1.890	405
Liquid fuels	261	314
Total	**252.173**	**93.282**

Source: Energy Portal (2021).

used 1,238 MWe (1.31 percent). The ratio of installed power using non-renewable resources to total installed power is 49.64 percent (Energy Portal, 2021).

According to resources, Turkey's electricity generation and installed power figures at the end of October 2020 are shown in Table 4.1.

4.4.1 GEOTHERMAL ENERGY GENERATION IN TURKEY

Geothermal energy is a renewable energy source obtained from hot fluid originating from the heat within the earth or from hot dry rock systems. Since geothermal energy

TABLE 4.2
Profile of Turkey's Geothermal Power Plants

Number of registered power plants	58
GPP installed power	1.515 MWe
Installed power ratio	1.63%
Annual electricity generation	~ 9.898 GWh
Ratio of generation to consumption	3.30%

Source: Energy Portal (2021); Energy Atlas (2021).

is a renewable energy source, its operating cost is cheaper than other energy types and its activation is faster (Takan & Kandemir, 2020). In many places, geothermal energy is used in daily life and healthcare, especially for electricity generation. Geothermal energy was economically developed in the 1900s for the first time in Italy, and has extensive potential in Turkey (Bekar, 2020).

Geothermal resources are classified according to water and steam temperatures. Geothermal resources with water and steam between 20°C and 70°C are called low-temperature geothermal resources, those that are 70–150°C are called medium-temperature geothermal resources, and those more than 150°C are called high-temperature geothermal resources (Takan & Kandemir, 2020). Geothermal resources are available, including varying temperatures, almost everywhere in Turkey. At low and medium temperatures, 90 percent of geothermal resources are directly suited for applications (heating, therapeutic tourism, various industrial applications, and so on), and 10 percent are indirect for electricity generation (Bekar, 2020). Table 4.2 illustrates Turkey's geothermal power plant profile.

A total of 7.626 GWh of electricity was produced from geothermal sources in Turkey at the end of October 2020. The ratio of electrical energy generated from geothermal sources to total generated electrical energy is 3.03 percent. The installed power of geothermal power plants is 1,515 MWe. The ratio of the power plants using other resources to the installed power is 1.63 percent (Energy Portal, 2021).

4.5   DATA AND ANALYSIS

In the study, also referred to as the energy exchange of Turkish Energy Market Operating Corporation (EPİAŞ), day-ahead electricity market's market clearing price data is used.

A day-ahead market (DAM) is an organized wholesale electricity market established and operated by EPİAŞ for electrical energy purchase and sale transactions on the basis of the settlement period to be delivered one day later. Long-term electricity purchases and sales in the electricity market are made through bilateral agreements. Complementing these agreements, DAM provides an environment for market participants to eliminate potential energy imbalances for the next day. The prices in the DAM are accepted as the electricity reference price (market clearing price

[MCP]) due to their proximity to real time. Although bilateral agreements make up the major volume of the electricity market, DAM is growing day by day.

The objectives of the day-ahead market are listed below (Atalay, 2009):

- To determine the electrical energy reference price
- To give market participants the chance to balance themselves by offering, in addition to their bilateral agreements, the ability to buy and sell energy for the next day
- To provide a day-ahead balanced system for the system operator
- To provide the system operator with the opportunity for day-ahead constraint management by creating bidding zones for large-scale and continuous constraints.

In this empirical study, a total of 1,097 days of EPİAŞ DAM MCP data from between November 10, 2017 and November 10, 2020 were used. MCP data are real price data in US dollars, obtained from the EPİAŞ Transparency Platform. Since there are separate spot prices for all 24 hours of the day, 24 different risky assets have been selected.

Mean, standard deviation, skewness, kurtosis, and median values were calculated for each risky asset. Statistical analysis results are shown in Table 4.3.

As can be seen in the correlation analysis, the correlation of each hour with the previous hour is quite high.

Hourly rates of return are designated by Liu and Wu (2007a), Gökgöz and Atmaca (2012, 2017), and deLlano-Paz et al. (2017) by using previously applied approaches. Koç & Şenel (2013) reported that unit costs of electricity generation based on plant types in Turkey and the unit energy generation cost of geothermal power plants were stated as \$33–40/MWh. In line with this information, the unit electricity generation cost of geothermal power plants was taken as \$35.6/MWh in this study. For the spot electricity market, the hourly rate of return is calculated as follows:

$$ret_{n,m} = \frac{\left(a_{n,m} - C_g\right)}{C_g}, (n = 1 \ldots, 24), (m = 1, \ldots, 1097) \qquad (4.44)$$

where $a_{n,m}$ represents n_{th} hour of m_{th} day day-ahead price, C_g represents the average generation cost for geothermal power plants, and $ret_{n,m}$ represents the hourly rate of return for the electric spot market (deLlano-Paz et al., 2017).

$$\overrightarrow{ret_1} = \begin{bmatrix} ret_{1,1} \\ ret_{1,2} \\ \vdots \\ \vdots \\ ret_{1,1097} \end{bmatrix}, \overrightarrow{r_2} = \begin{bmatrix} ret_{2,1} \\ ret_{2,2} \\ \vdots \\ \vdots \\ ret_{3,1097} \end{bmatrix}, \overrightarrow{r_3} = \begin{bmatrix} ret_{3,1} \\ ret_{3,2} \\ \vdots \\ \vdots \\ ret_{3,1097} \end{bmatrix} \ldots\ldots\ldots, \overrightarrow{r_{24}} = \begin{bmatrix} ret_{24,1} \\ ret_{24,2} \\ \vdots \\ \vdots \\ ret_{24,1097} \end{bmatrix} \qquad (4.45)$$

where $\overrightarrow{ret_n}[n = 1, \ldots, 24]$ represents the vector of the rate of return for each hour.

TABLE 4.3
Statistical Analysis of 24 Risky Assets

Hour	Mean $/MWh	Std. deviation	Skewness	Kurtosis	Median $/MWh	Day count
0	43.92	11.74	−1,415	2.280	46.24	1,097
1	44.95	9.97	−1,488	3.487	46.71	1,097
2	41.13	10.76	−0,949	1.326	42.66	1,097
3	37.80	11.93	−1,007	1.228	39.24	1,097
4	35.86	12.40	−0,925	0.870	37.38	1,097
5	36.34	12.65	−0,939	0.881	37.68	1,097
6	36.90	14.25	−0,892	0.457	38.18	1,097
7	40.41	14.25	−1,149	0.924	42.89	1,097
8	45.66	13.33	−1,659	2.843	49.58	1,097
9	45.71	14.99	−1,648	2.186	51.77	1,097
10	46.58	13.58	−1,772	3.166	51.54	1,097
11	47.89	12.64	−1,970	4.351	52.49	1,097
12	43.06	13.49	−1,351	1.755	45.5	1,097
13	44.18	13.39	−1,479	2.232	47.02	1,097
14	46.84	12.62	−1,715	3.869	49.96	1,097
15	46.34	13.20	−1,448	3.343	49.46	1,097
16	47.52	12.81	−1,22	5.170	50.53	1,097
17	48.55	11.49	−1,213	6.109	51.07	1,097
18	49.52	9.99	−1,499	4.501	52.69	1,097
19	50.81	7.96	−0,673	5.865	53.09	1,097
20	51.05	6.59	−0,857	1.150	53.07	1,097
21	50.18	6.71	−0,861	1.199	51.67	1,097
22	47.44	8.15	−1,059	1.932	49.09	1,097
23	43.87	10.13	−1,196	2.093	45.05	1,097

After performing statistical analysis, correlation analysis was performed for 24 risky assets. Table 4.4 shows the results of the correlation analysis.

Unlike the finance literature, as seen in the above equations, the rate of return on electric spot markets is determined using actual selling prices and production costs. The average return rate, standard deviation, and covariance matrix for each hour are calculated considering Gökgöz and Atmaca (2012) as follows, respectively.

$$\overline{ret}_n = \frac{1}{1097}\left(\sum_{m=1}^{1097} ret_{n,m}\right) \tag{4.46}$$

$$\sigma_n = \sqrt{\frac{1}{1097}\sum_{m=1}^{1097}\left(ret_{n,m} - \overline{ret}_n\right)^2} \tag{4.47}$$

$$\sigma_{x,y} = \frac{1}{1097}\sum_{m=1}^{1097}\left(ret_{x,m} - \overline{ret}_x\right)\left(ret_{y,m} - \overline{ret}_y\right) \tag{4.48}$$

TABLE 4.4
Results of the Correlation Analysis of 24 Risky Assets

Hour	0	1	2	3	4	5	6	7	8	9	10	11	12	13	14	15	16	17	18	19	20	21	22	23
0	1																							
1	0.778	1																						
2	0.61	0.778	1																					
3	0.681	0.732	0.803	1																				
4	0.621	0.663	0.721	0.819	1																			
5	0.656	0.688	0.675	0.708	0.78	1																		
6	0.574	0.584	0.476	0.466	0.437	0.754	1																	
7	0.548	0.543	0.499	0.492	0.423	0.633	0.79	1																
8	0.396	0.367	0.374	0.342	0.301	0.465	0.584	0.768	1															
9	0.436	0.337	0.341	0.327	0.292	0.474	0.61	0.748	0.888	1														
10	0.471	0.394	0.385	0.366	0.327	0.491	0.634	0.782	0.872	0.935	1													
11	0.47	0.413	0.404	0.371	0.348	0.494	0.617	0.751	0.829	0.876	0.945	1												
12	0.576	0.494	0.471	0.476	0.441	0.576	0.66	0.785	0.745	0.804	0.86	0.868	1											
13	0.595	0.501	0.487	0.5	0.476	0.59	0.618	0.713	0.69	0.755	0.802	0.816	0.902	1										
14	0.505	0.428	0.44	0.443	0.416	0.529	0.594	0.735	0.774	0.833	0.868	0.883	0.874	0.898	1									
15	0.511	0.428	0.449	0.445	0.432	0.538	0.578	0.704	0.714	0.787	0.808	0.824	0.837	0.861	0.959	1								
16	0.517	0.443	0.444	0.438	0.427	0.536	0.567	0.691	0.69	0.741	0.774	0.797	0.79	0.801	0.889	0.914	1							
17	0.463	0.435	0.427	0.39	0.355	0.504	0.567	0.675	0.657	0.683	0.735	0.776	0.745	0.727	0.807	0.825	0.913	1						
18	0.488	0.5	0.445	0.408	0.348	0.508	0.609	0.652	0.594	0.622	0.691	0.738	0.705	0.682	0.713	0.712	0.749	0.855	1					
19	0.366	0.441	0.367	0.3	0.241	0.418	0.517	0.48	0.435	0.43	0.483	0.534	0.484	0.461	0.486	0.486	0.556	0.692	0.79	1				
20	0.43	0.519	0.414	0.317	0.273	0.452	0.503	0.475	0.425	0.406	0.457	0.511	0.465	0.452	0.455	0.451	0.487	0.594	0.68	0.788	1			
21	0.51	0.593	0.504	0.404	0.347	0.474	0.51	0.514	0.45	0.435	0.49	0.536	0.524	0.518	0.498	0.485	0.515	0.584	0.669	0.679	0.833	1		
22	0.506	0.622	0.572	0.468	0.405	0.491	0.501	0.546	0.461	0.423	0.502	0.555	0.555	0.527	0.514	0.502	0.536	0.608	0.68	0.644	0.745	0.852	1	
23	0.553	0.632	0.581	0.509	0.486	0.553	0.509	0.572	0.448	0.427	0.493	0.523	0.576	0.572	0.549	0.562	0.588	0.623	0.647	0.556	0.627	0.726	0.848	1

$$\text{Covariance Matrix} = \begin{bmatrix} \sigma_1^2 & \sigma_{1,2} & \cdots & \sigma_{1,24} \\ \sigma_{2,1} & \sigma_2^2 & & \\ \vdots & & \ddots & \\ \sigma_{24,1} & & & \sigma_{24}^2 \end{bmatrix} \qquad (4.49)$$

In the study, 24-hour day-ahead market USD-based market clearing prices were used for a total of 1,097 days between November 10, 2017 and November 10, 2020, and optimization studies were carried out with MV and MAD methods on a total of 26,328 pieces of data. By using geothermal power plant generation costs, the objective functions of minimum risk, maximum return, maximum Sharpe ratio, and maximum utility ($A = 3$, $A = 4$, $A = 5$) were applied for both models, and optimal portfolios were calculated. The performances of the optimal portfolios were measured by the Sharpe ratio. MV and MAD optimization results were compared according to their portfolio performances, and recommendations were made to electricity market decision makers for their investments.

The credentials of the empirical study are shown in Table 4.5.

The assumptions in the study are as follows:

- The unit electricity generation cost of the geothermal power plant was taken as $35.6/MWh.
- Electricity can be traded for seven days, on both weekdays and weekends.
- Electricity can be traded 24 hours a day.
- Investors' risk aversion levels are taken as $A = 3$, 4, and 5.
- There is no power outage.

On the other side, the amount of electricity supplied by the investor to the market does not affect the price of the system. Electricity production cost is fixed, and there is no problem or congestion in electricity transmission. Bids can be split into an unlimited

---

TABLE 4.5
Credentials of Empirical Studies

Topic	MV optimization model	MAD optimization model
Objective functions	Minimum Risk, Maximum Return, Maximum Sharpe Ratio, Maximum Utility ($A = 3, A = 4, A = 5$)	
Power plant type	Geothermal Power Plant	
Electricity generation cost	35.6 $/MWh	
Investment period	Weekdays and Weekend	
Weekdays	5 days (Monday, Tuesday, Wednesday, Thursday, Friday)	
Weekend	2 days (Saturday, Sunday)	
Market data	EPİAŞ Day Ahead Market (10.11.2017–10.11.2020) 1,097 days' data	
Count of risky assets	24 (hours of day)	

TABLE 4.6
Returns and Standard Deviations of 24 Risky Assets based on Generation Cost

Hours	Return (%)	Std. Deviation (%)	Hours	Return (%)	Std. Deviation (%)
0	23.37	32.98	12	20.96	37.90
1	26.26	28.01	13	24.12	37.61
2	15.55	30.23	14	31.6	35.43
3	6.18	33.51	15	30.19	37.06
4	0.75	34.83	16	33.48	35.99
5	2.09	35.54	17	36.4	32.26
6	3.66	40.03	18	39.13	28.06
7	13.53	40.02	19	42.73	22.36
8	28.26	37.43	20	43.41	18.52
9	28.42	42.09	21	40.98	18.85
10	30.85	38.15	22	33.26	22.88
11	34.53	35.51	23	23.24	28.46

number of smaller pieces. All bids shall be received by the market. A less risky portfolio with the same return level and the highest return portfolio with the same risk level are preferred by the investors (Gökgöz & Atmaca, 2017).

By using geothermal power plant generation cost, the average return rate and standard deviations are calculated for 24 risky assets as follows by using the formulas specified in Equations 4.46 and 4.47.

As seen in Table 4.6, when risky assets are evaluated on a return basis, the minimum return asset is calculated as the 4th hour and the maximum return asset is calculated as the 20th hour. When it comes to risk, the minimum risk asset is calculated as the 20th hour, and the maximum risk asset as the 9th hour. The asset with minimum risk and maximum return is 20 hours. When 24 risky assets are evaluated on the basis of risk and return, the two most efficient assets are at the 20th and 21st hours.

MV and MAD models were created and optimizations were performed for both models and objective functions using 24 risky assets. The efficient frontier graphics obtained are shown in Figures 4.4 and 4.5.

When Figure 4.4 was examined, results in accordance with the literature were obtained for six calculated portfolios. All the calculated portfolios are shown on efficient frontiers. The risk value of the minimum risk portfolio is calculated as 17.49 percent. The risks of maximum utility portfolios decrease as the risk aversion level increases. The risk value for $A = 3$ was calculated as 17.96 percent, 17.91 percent for $A = 4$, and 17.89 percent for $A = 5$. The return rate is calculated as 43.41 percent for the maximum return portfolio. The Sharpe ratio of the maximum Sharpe portfolio is calculated as 2.3789.

When Figure 4.5 was examined, the results in accordance with the literature were obtained for six calculated portfolios. All the calculated portfolios are shown on efficient frontiers. The minimum risk portfolio's risk value is calculated as 14.26 percent. The risks of maximum utility portfolios decrease as the risk aversion level increases.

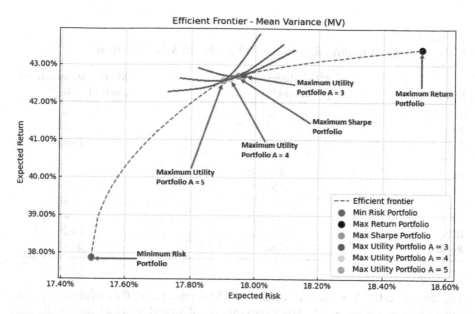

FIGURE 4.4   MV efficient frontier chart and optimal portfolios.

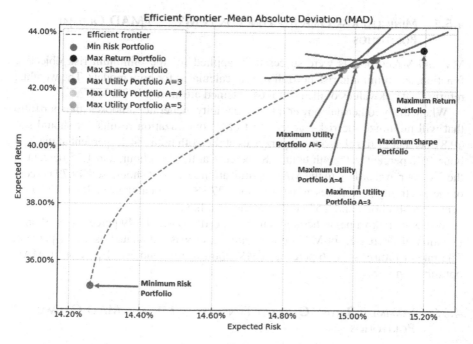

FIGURE 4.5   MAD efficient frontier chart and optimal portfolios.

TABLE 4.7
MV and MAD Minimum Risk Portfolio Weight, Risk, and Return

Hour	MV—Minimum risk portfolio (%)	MAD—Minimum risk portfolio (%)
0–2	—	—
3	1.75	—
4	9.46	17.80
5–18	—	—
19	4.86	4.53
20	49.81	49.79
21	34.12	27.88
22–23	—	—
Risk	17.49	14.26
Return	37.86	35.11

The risk value for $A = 3$ was calculated as 15.05 percent, 15.00 percent for $A = 4$, and 14.97 percent for $A = 5$. The return rate is calculated as 43.41 percent for the maximum return portfolio. The maximum Sharpe ratio portfolio's Sharpe ratio is calculated as 2.8612.

4.5.1 MINIMUM RISK OBJECTIVE FUNCTION IN MV AND MAD OPTIMAL PORTFOLIOS

MV and MAD models were successfully applied for the minimum risk objective function, and minimum risk portfolios were calculated for both models. The weights, returns, and risk values of the portfolios obtained are shown in Table 4.7.

When the geothermal power plant electricity generator chooses the portfolio that will minimize the risk—according to MV optimization results—it should sell 49.81 percent of the electricity it generates at the 20th hour, 34.12 percent at the 21st hour, 9.46 percent at the 4th hour, 4.86 percent at the 19th hour, and 1.75 percent at the 3rd hour. According to the MAD optimization results, it should sell 49.79 percent of the electricity it produces at the 20th hour, 27.88 percent at the 21st hour, 17.8 percent at the 4th hour, and 4.53 percent at the 19th hour.

While the risk value of the minimum risk portfolio was 17.49 percent in MV optimization, it decreased in MAD optimization and was calculated as 14.26 percent. The rate of return is 37.86 percent in MV optimization and 35.11 percent in MAD optimization.

4.5.2 MAXIMUM RETURN OBJECTIVE FUNCTION IN MV AND MAD OPTIMAL PORTFOLIOS

For the maximum return objective function, MV and MAD models have been successfully applied and maximum return portfolios have been calculated for both models. The weights, return, and risk values of the portfolios obtained are shown in Table 4.8.

TABLE 4.8
MV and MAD Maximum Return Portfolio Weight, Risk, and Return

Hour	MV—Maximum return portfolio (%)	MAD—Maximum return portfolio (%)
0–19	—	—
20	100.00	100.00
21–23	—	—
Risk	18.52	15.19
Return	43.41	43.41

TABLE 4.9
MV and MAD Maximum Sharpe Ratio Portfolio Weight, Risk, and Return

Hour	MV—Maximum Sharpe ratio portfolio (%)	MAD—Maximum Sharpe ratio portfolio (%)
0–18	—	—
19	4.32	2.36
20	67.16	84.93
21	28.52	12.70
22–23	—	—
Risk	17.94	15.06
Return	42.69	43.09

When the geothermal power plant electricity producer chooses the portfolio that will maximize its return, a 43.41 percent return rate is expected, according to both the MV and MAD optimization results. The risk value of the maximum return portfolio was found as 18.52 percent in MV optimization and 15.19 percent in MAD optimization.

The geothermal power plant electricity producer is required to sell 100 percent of the electricity it produces in the 20th hour, according to both MV and MAD optimization results. Table 4.4 shows that the minimum risky and maximum return asset among 24 risky assets is the 20th hour. Therefore, the maximum return portfolio obtained as a result of the optimization made with both models consists of only the 20th hour.

4.5.3 MAXIMUM SHARPE RATIO OBJECTIVE FUNCTION IN MV AND MAD OPTIMAL PORTFOLIOS

For the maximum Sharpe ratio objective function, MV and MAD models have been successfully applied, and portfolios with maximum Sharpe ratios have been calculated for both models. The weights, return, and risk values of the portfolios obtained are shown in Table 4.9.

When the geothermal power plant electricity producer chooses the portfolio that will maximize the Sharpe ratio, it is seen that it should invest in the 19th, 20th, and 21th hour, according to both MV and MAD optimization results. According to the MV optimization results, it is required to sell 67.16 percent of the electricity it produces at the 20th hour, 28.52 percent at the 21st hour, and 4.32 percent at the 19th hour. According to the MAD optimization results, it should sell 84.93 percent of the electricity it produces at the 20th hour, 12.70 percent at the 21st hour, and 2.36 percent at the 19th hour.

When the risks and returns of the optimal portfolios obtained as a result of MV and MAD optimizations are compared, the portfolio obtained as a result of MAD optimization has both low risk and high return.

4.5.4 MAXIMUM UTILITY ($A = 3$, $A = 4$, $A = 5$) OBJECTIVE FUNCTION IN MV AND MAD OPTIMAL PORTFOLIOS

MV and MAD models were successfully applied for the maximum utility ($A = 3$, $A = 4$, $A = 5$) objective function, and the maximum utility ($A = 3$, $A = 4$, $A = 5$) portfolios were calculated for both models. The weights, return, and risk values of the portfolios obtained are shown in Table 4.10.

A total of six optimizations were made for three different risk aversion levels ($A = 3$, $A = 4$, $A = 5$) with MV and MAD optimizations. In general, if the results are evaluated and the risk aversion coefficient increases, the risk value of the optimal portfolio obtained decreases and, accordingly, the rate of return decreases.

When the optimization results were examined, the risk values were lower than the MV model in all of the MAD model results. The rate of return in all the MAD model results, however, was higher than the MV model.

When the geothermal power plant electricity producer chooses the portfolio that will maximize the utility for three different risk aversion levels ($A = 3$, $A = 4$, $A = 5$), it is seen that it should invest in the 19th, 20th, and 21th hours, according to both MV and MAD optimization results.

According to the results of MV optimization, the electricity produced when $A = 3$ should be sold for 68.96 percent at the 20th hour, 26.90 percent at the 21st hour,

---

TABLE 4.10
MV and MAD Maximum Utility ($A = 3$, $A = 4$, $A = 5$) Portfolio Weight, Risk, and Return

Hour	MV—Maximum utility risk portfolios (%) $A = 3$, $A = 4$, $A = 5$			MAD—Maximum utility risk portfolios (%) $A = 3$, $A = 4$, $A = 5$		
0–18	—	—	—	—	—	—
19	4.14	4.61	4.90	2.54	1.45	2.36
20	68.96	64.20	61.35	83.65	77.23	70.36
21	26.90	31.18	33.75	13.81	21.32	27.28
22 - 23	—	—	—	—	—	—
Risk	17.96	17.91	17.89	15.05	15.00	14.97
Return	42.73	42.62	42.55	43.06	42.89	42.73

and 4.14 percent at the 19th hour, $A = 4$ should be 64.20 percent at the 20th hour, 31.18 percent at the 21st hour, and 4.61 percent at the 19th hour, and, for $A = 5$, it's 61.35 percent at the 20th hour, 33.75 percent at the 21st hour, and 4.90 percent at the 19th hour.

According to the MAD optimization results, the electricity produced when $A = 3$ should be sold for 83.65 percent at the 20th hour, 13.81 percent at the 21st hour, and 2.54 percent at the 19th hour; when $A = 4$, electricity should be sold at 77.23 percent at the 20th hour, 21.32 percent at the 21st hour, and 1.45 percent at the 19th hour, and, for $A = 5$, it's 70.36 percent at the 20th hour, 27.28 percent at the 21st hour, and 2.36 percent at the 19th hour.

4.5.5 COMPARISON OF PERFORMANCES OF THE MV AND MAD OPTIMAL PORTFOLIOS

Portfolio performance can be measured with more than one performance measurement approach. These different approaches can be used individually or together according to the needs of the investor. The Sharpe ratio is one of the most well-known and frequently used performance indicators in portfolio theory and is also called reward to variability (RVAR). Risk and return are used as parameters in performance measurement (Israelsen, 2005; Gökgöz, 2009; Cucchiella et al., 2016). The Sharpe ratio is calculated as follows.

$$RVAR_p = \frac{ar_p - ar_f}{\sigma_p} \qquad (4.50)$$

"$RVAR_p$" represents the portfolio's Sharpe ratio, "ar_p" represents the portfolio's return, "ar_f" represents the risk-free asset's return, "σ_p" represents the risk of the portfolio. Twenty-four risky assets were used in the study, but risk-free assets (bilateral contracts) were not included. Therefore, ar_f is taken "0" in the calculations.

In the study, Sharpe ratios of 12 optimal portfolios obtained as a result of MV and MAD optimizations were calculated; these are shown in Table 4.11.

Portfolio performance comparisons according to Table 4.11 are as follows:

- Among MV and MAD optimizations, the portfolio with the highest performance is the maximum Sharpe ratio portfolio, as expected.
- Among the MV and MAD optimizations, the lowest performance portfolio is the minimum risk portfolio.
- As the risk aversion coefficient (A) increases in both MV and MAD optimizations, portfolio performances decrease.
- MAD optimal portfolios performed better than MV optimal portfolios. For the six objective functions shown in Table 4.11, the average Sharpe ratio performance of optimal portfolios is calculated as 2.3370 for MV and 2.7925 for MAD. As a result, MAD optimal portfolios are more efficient than the MV model.

TABLE 4.11
The Sharpe Ratios of Optimal Portfolios

Objective functions	MV optimization	MAD optimization
Minimum risk portfolio	2.1640	2.4620
Maximum return portfolio	2.3437	2.8571
Maximum Sharpe ratio portfolio	2.3789	2.8612
Maximum utility portfolio $A = 3$	2.3788	2.8611
Maximum utility portfolio $A = 4$	2.3786	2.8586
Maximum utility portfolio $A = 5$	2.3777	2.8548
Average Sharpe ratio	2.3370	2.7925

4.6 CONCLUSION

Electricity has a very important role to play in every stage of human life. Population growth and technological developments increase energy consumption every day. Electricity is produced using both renewable and non-renewable energy sources. Electricity should be used safely and efficiently since all forms are limited and the produced electricity cannot be stored. Electricity producers operating in the electricity market should consider market risks and be able to better manage their risks. For this, they should optimize their market bid strategies and generation capacities. The use of renewable energy resources is encouraged by the governments because the non-renewable energy resources will be depleted soon, causing environmental problems and external dependence. Therefore, in this study, geothermal power plants generating electricity with geothermal resources—a type of renewable energy—were used, and the importance of renewable energy was emphasized.

In addition, for electricity producers to provide the best bidding strategy in Turkish electricity market environment, MV and MAD optimization models have been applied. These two models have found wide application in finance literature. Its implementation in the electricity market, however, is quite new. A total of 12 optimizations were made for six different objective functions (minimum risk, maximum return, maximum Sharpe ratio, maximum utility $A = 3$, $A = 4$, $A = 5$) with MV and MAD models.

The findings of the empirical study are listed below:

- As a result of optimizations, optimal portfolios were successfully produced and compared. Weights, risks, and returns of all optimal portfolios are presented in the study. It has been observed that the optimal solutions vary according to the purpose of the investor or decision maker in each model. Optimal portfolios for different levels of risk aversion were calculated and the results evaluated. Portfolio performances were calculated for both models, and it was determined that MAD optimal portfolios gave better results. The results of different optimization models for various objectives are shown. These results are thought to contribute to the investor's decision-making process or the decision maker.

- In the empirical study, optimizations were carried out with MV and MAD models by using geothermal power plant generation cost. These models can be applied, using different generation costs, to other types of electricity generation. In this way, it can be analyzed to determine how electricity producers' bidding strategies have changed.
- Additionally, Kasenbacher et al. (2017) applied the MV and MAD optimization models to the financial market, comparing their portfolio performances. The application result was in line with our study results, and it was stated that the MAD optimal portfolios gave better results. In our study, it has been shown that the application of the Sharpe ratio as well as MV and MAD optimization models to the financial and electricity market yields similar results.

As a result, it has been shown that the MV and MAD optimization models, which are frequently used in financial markets, can also be applied to electricity markets. As a portfolio optimization approach for electricity markets, the findings of the study will make substantial contributions to the literature in this area. It will also provide valuable empirical results for electricity market decision makers in setting up decision support systems and determining investment strategies.

REFERENCES

Ağır, H., Özbek, S., & Türkmen, S. (2020). Determinants of renewable energy sources in Turkey: An empirical estimation. *International Journal of Economic Studies*, 6(4), 39–48.

Atalay, H. (2019). Day ahead market. Retrieved from www.dunyaenerji.org.tr/wp-content/uploads/2019/01/GOPOcak2019.pdf Access date: December 22, 2020

Bekar, N. (2020). Turkey's energy geopolitics in terms of renewable energy sources. *Turkish Journal of Political Science*, 3(1), 37–54.

Bhattacharya, A., & Kojima, S. (2012). Power sector investment risk and renewable energy: A Japanese case study using portfolio risk optimization method. *Energy Policy, 40*, 69–80.

Boroumand, R. H., Goutte, S., Porcher, S., & Porcher, T. (2015). Hedging strategies in energy markets: The case of electricity retailers. *Energy Economics, 51*, 503–509.

BOUN (2021). Bogazici University center for climate change and policy studies. Retrieved from: http://climatechange.boun.edu.tr/iklim-degisikligi-ve-yenilenebilir-enerji/ Access date: March 1, 2021

Byström, H. N. (2003). The hedging performance of electricity futures on the Nordic power exchange. *Applied Economics, 35*(1), 1–11.

Copeland T. E., Weston J. F., Shastri, K. (2005). Financial theory and corporate policy. 4th edition. Columbus: Addison Wesley Pearson.

Cucchiella, F., D'Adamo, I., & Gastaldi, M. (2016). Optimizing plant size in the planning of renewable energy portfolios. *Letters in Spatial and Resource Sciences*, 9(2), 169–187.

Dahlgren, R., Liu, C. C., & Lawarree, J. (2003). Risk assessment in energy trading. *IEEE Transactions on Power Systems*, 18(2), 503–511.

deLlano-Paz, F., Calvo-Silvosa, A., Antelo, S. I., & Soares, I. (2017). Energy planning and modern portfolio theory: A review. *Renewable and Sustainable Energy Reviews, 77*, 636–651.

Defusco, R. A., McLeavey, D. W., Pinto, J. E., & Runkle, D. E. (2004). Quantitative investment analysis. Hoboken: John Wiley.

Energy Atlas (2021). Retrieved from: www.enerjiatlasi.com/jeotermal/ Access date: January 22, 2021.

Energy Portal (2021). Retrieved from: www.enerjiportali.com/turkiye-elektrik-enerjisi-uretim-istatistikleri-ekim-2020/ Access date: January 21, 2021.

Feng, D., Gan, D., Zhong, J., & Ni, Y. (2007). Supplier asset allocation in a pool-based electricity market. *IEEE Transactions on Power Systems, 22*(3), 1129–1138.

Gökgöz, F. (2009). Mean variance optimization via factor models in the emerging markets: Evidence on the Istanbul stock exchange. *Investment Management and Financial Innovations, 6*(3), 43–53.

Gökgöz, F., & Atmaca M. E. (2012). Financial optimization in the Turkish electricity market: Markowitz's mean-variance approach. *Renewable and Sustainable Energy Reviews, 16*(1), 357–368.

Gökgöz, F., & Atmaca M. E. (2017). Portfolio optimization under lower partial moments in emerging electricity markets: Evidence from Turkey. *Renewable and Sustainable Energy Reviews, 67*, 437–449.

Grootveld, H., & Hallerbach, W. (1999). Variance vs downside risk: Is there really that much difference? *European Journal of Operational Research, 114*(2), 304–319.

Huang, Y. H., & Wu, J. H. (2016). A portfolio theory based optimization model for steam coal purchasing strategy: A case study of Taiwan Power Company. *Journal of Purchasing and Supply Management, 22*(2), 131–140.

Israelsen, C. L. (2005). A refinement to the Sharpe ratio and information ratio. *Journal of Asset Management, 5*(6), 423–427.

Ivanova, M., & Dospatliev, L. (2017). Application of Markowitz portfolio optimization on Bulgarian stock market from 2013 to 2016. *International Journal of Pure and Applied Mathematics, 117*(2), 291–307.

Kardiyen, F. (2007). A study on portfolio optimization with linear programming and its application to IMKB data. *Atatürk University Journal of Economics and Administrative Science, 21*(2), 15–28.

Kardiyen, A. G. F. (2008). The use of mean absolute deviation model and Markowitz model in portfolio optimization and its application to IMKB data. *The Journal of Faculty of Economics Administrative Sciences, 13*(2), 335–350.

Kasenbacher, G., Lee, J., & Euchukanonchai, K. (2017). Mean-variance vs. mean-absolute deviation: A performance comparison of portfolio optimization models. University of British Columbia.

Kayışoğlu, B., & Diken, B. (2019). Current status and problems of renewable energy usage in Turkey. *Journal of Agricultural Machinery Science, 15*(2), 61–65.

Kazempour, S. J., & Moghaddam, M. P. (2011). Risk-constrained self-scheduling of a fuel and emission constrained power producer using rolling window procedure. *International Journal of Electrical Power & Energy Systems, 33*(2), 359–368.

Koç, E., & Şenel, M. C. (2013). Turkey's energy potential and investment-cost analysis. *The Journal of Thermodynamics, 245*, 72–84.

Konno, H., & Koshizuka, T. (2005). Mean-absolute deviation model. *IIE Transactions, 37*(10), 893–900.

Konno, H., & Yamazaki, H. (1991). Mean-absolute deviation portfolio optimization model and its applications to Tokyo stock market. *Management Science, 37*(5), 519–531.

Liu, M. (2004). Energy allocation with risk management in electricity markets (PhD dissertation). Hong Kong: Department of Electrical and Electronical Engineering, The University of Hong Kong.

Liu, M., & Wu, F. F. (2006). Managing price risk in a multimarket environment. *IEEE Transactions on Power Systems, 21*(4), 1512–1519.

Liu, M., & Wu, F. F. (2007a). Portfolio optimization in electricity markets. *Electric Power Systems Research*, 77(8), 1000–1009.

Liu, M., & Wu, F. F. (2007b). Risk management in a competitive electricity market. *International Journal of Electrical Power & Energy Systems*, 29(9), 690–697.

Markowitz H. (1952). Portfolio selection. *The Journal of Finance*, 7(1), 77–91.

Mansini R., Ogryczaki W., & Speranza M. G. (2003). LP solvable models for portfolio optimization: A classification and computational comparison, *IMA Journal of Management Mathematics*, 14, 187–220.

Munoz, J. I., de la Nieta, A. A. S., Contreras, J., & Bernal-Agustín, J. L. (2009). Optimal investment portfolio in renewable energy: The Spanish case. *Energy Policy*, 37(12), 5273–5284.

Öztürk, H. K., Yilanci, A., & Atalay, O. (2007). Past, present and future status of electricity in Turkey and the share of energy sources. *Renewable and Sustainable Energy Reviews*, 11(2), 183–209.

Pindoriya, N. M., Singh, S. N., & Singh, S. K. (2010). Multi-objective mean–variance–skewness model for generation portfolio allocation in electricity markets. *Electric Power Systems Research*, 80(10), 1314–1321.

Şenocak, F., & Kahveci, H. (2016). Periodic price averages forecasting of MCP in day-ahead market. In *2016 National Conference on Electrical, Electronics and Biomedical Engineering (ELECO)* (pp. 664–668). IEEE.

Simaan, Y. (1997). Estimation risk in portfolio selection: The mean variance model versus the mean absolute deviation model. *Management Science*, 43(10), 1437–1446.

Statman, M. (1987). How many stocks make a diversified portfolio? *Journal of Financial and Quantitative Analysis*, 22, 353–363.

Suksonghong, K., Boonlong, K., & Goh, K. L. (2014). Multi-objective genetic algorithms for solving portfolio optimization problems in the electricity market. *International Journal of Electrical Power & Energy Systems*, 58, 150–159.

Takan, M. A., & Kandemir, S. Y. (2020). Primary energy supply for the evaluation of geothermal energy in Turkey. *European Journal of Science & Technology*, 20, 381–385.

Wattoo, W. A., Kaloi, G. S., Yousif, M., Baloch, M. H., Zardari, B. A., Arshad, J., Farid, G., Younas, T., & Tahir, S. (2020). An optimal asset allocation strategy for suppliers paying carbon tax in the competitive electricity market. *Journal of Electrical Engineering & Technology*, 15(1), 193–203.

Zhu, H., Wang, Y., Wang, K., & Chen, Y. (2011). Particle swarm optimization (PSO) for the constrained portfolio optimization problem. *Expert Systems with Applications*, 38(8), 10161–10169.

5 The Impact of Traditional Natural Stones on Energy Efficiency for Sustainable Architecture

The Case of an Authentic Restaurant in Harput Region

F. Balo and H. Polat

DOI: 10.1201/9781003240129-5

NOMENCLATURE

BIM   Building information modeling
BEM   Building energy model
C_{th} Thermal mass (kJ/K)
EUI   Energy use intensity
HVAC   Heating, ventilating and air conditioning
LCA   Life cycle analysis
LCC   Life cycle cost
LEED   Leadership in energy and environmental design
U   Heat transfer coefficient (W/m²K)

5.1 INTRODUCTION

The construction industry, accounting for about 35–40% of energy utilization world-wide, is one of the main power expending industries [1,2].

In buildings, energy analysis is essential for ascertaining their energy performance. Investigations have been done to examine the energy efficiency of constructions and to improve analysis methods. Computer software analyses are efficient for a structure's energy investigation. The next trend of energy performance norms will be based on analysis methods. For this purpose, the energy and thermal performance of energy-efficient materials for varying ecological conditions require to be assessed [3]. In the building design, a broadly accepted opinion is that using efficiency feedback analysis will result in developing efficiency in the plan. Another view is that integration and automation of efficiency simulation into the plan in the early stages will also help in crafting greater performing plans. These opinions have given rise to the researchers proposing the concept of "designing-in efficiency." An efficient building design depends on the validated direct feedback among the fields of the plan, energy optimization, and simulation presented during the first steps of the plan phase where it has been known that such a decision-making has the highest effect on the overall construction efficiency [4]. In other words, efficiency plays a substantial role in strengthening the plan, as it helps in decision-making with the help of computational tools for the construction plan [5]. Nonetheless, plan experts are mostly unable to efficiently design plans and analyze their effect on energy expenditure owing to an array of challenges between energy efficiency fields and design [6–8]. For this reason, the ideal solution to overcome these obstacles is the simulation programs. These programs can ensure a fast generation of plan options, rapid assessment of plan options, trade-off simulation for antagonist criteria, and a research methodology to define plan options with optimum suitable efficiency.

Besides this, the national policy of many governments is to decrease CO_2 emissions and also generation of energy from the use of fossil-based fuels. In tandem with greater expectations about comparison with the total warmth of the indoor ambience, TSE 825 [9] directive has led to novel norms and a kind of building structure planned in such a way that energy expenditure by a building during its service life is minimized. Optimizing energy efficiency must be taken into consideration with building components and the energy approach of the construction for further effective utilization, along with developing energy efficiency [10].

The external wall layers are the functional dividers between the unconditioned and conditioned environment of the construction involving the resistance to air, water, light, noise, and heat transfer. The authentic building is a building that is constructed in accordance with the local exterior architectural appearance of the historical region. The thermal insulation of the external wall, which is a section of the external wall layers, should be avoided since the authentic appearance is important in historical building design [11]. The heating and cooling of authentic buildings without proper thermal insulation requirements increases energy expenditure for maintaining and using the building. It also gives rise to growing adverse ecological effect. The presence of a building's cold areas, such as attics and walls, which are inadequately insulated or uninsulated, lead to overheating in summer and constructional deficiencies owing to condensation within the construction or increased thermal loss in the winter. Poor condensing control affects the efficiency of construction components and damages the lifetime and the durability of the construction itself. Condensing may generate presence of the mold allergens, which gives rise to unhealthy working and living environments and chronic health troubles. For these reasons, the purpose must be how to develop the energy quality of the architectural ambience while sustaining the authentic texture of buildings with a traditional appearance.

While working on an authentic construction, there are energy-based issues that the building experts and users deal with, like architecture type of a construction and certain parameters of conventional construction materials. To fulfill the necessities of the TSE 825 directive, the utilization of natural stone as the primary construction material is important as a sustainable alternative. For all building materials, TSE 825 determines the thermal insulation requirements in buildings. According to TSE 825, insulation is supposed to be breathable. The breathable building wall is one that allows odors, toxins in the air, and moisture, to move from indoors of the construction to outdoors.

Natural stone is a traditional material obtained from natural reserves called quarry. After a basic technique of natural stone cutting, the stone is converted into a structural material. Natural stone requires minimum upkeep. Structure stones were broadly utilized for building vaults, domes, walls, or massive foundations in the Middle Ages. Various forms of natural stone were utilized for interior design, the indoor flooring of buildings as well as for decorative facade ornaments [12]. The utilization of traditional natural stone material sustains the facades' authentic value. The natural stone material has a direct effect on the bioclimatic construction properties and aesthetical appearance of authentic buildings. The exploitation of traditional natural stone from these quarries contributes to the authentic appearance of new architectural buildings. It also prevents negative potencies regarding endurance and permanence, therefore, it enables the conservation of architectural value. A successfully designed authentic architecture sample raises the value of the property and may also give support to national and international tourism [13].

Bioclimatic construction refers to the plans of spaces (exterior and interior) and buildings with local acclimatization, for the purpose of supplying visual, thermal comfort, and utilizing solar energy and other nature-based resources. Bioclimatic plan takes into consideration the regional climate and includes some basis. For example.

- The thermal conservation of the constructions in summer months and in winter months is particularly used for providing sufficient air tightness. This is done by utilizing natural materials that are added to the building envelope.
- The protection of buildings from the sun in the summer months is done mainly through shading, and also with the building envelope's suitable configuration (such as utilization of reflective natural surfaces and proper colors).
- The improvement of buildings' interior acclimatization such as cool storage or heat storage in the walls is done with the help of the natural materials [14].

Many modern buildings have not taken advantage of the passive configuration schemes, which are seen in conventional buildings. Conventional constructions are generally known as "breathable structures." In conventional buildings, local materials and forms are utilized widely at a specific place. In some cases, these materials are determined by an emigrant population's potent ethnic impact (generally unpretentious, modest, and as a mix of traditional and more contemporary styles; or a hybrid of a few forms). Conventional buildings are frequently owner-built structures built by persons who know regional materials, local climate circumstances, and regional construction techniques and customs, as defined under public architecture. Hence, the primary interest of this research is to develop to the thermal performance of newly designed authentic buildings using natural materials. The energy configuration schemes in conventional buildings are significant. Therefore, the energy performance is analyzed utilizing REVIT program, which has the BIM tool. BIM has been determined as a beneficial software that supplies timely data nearly throughout overall structure's life and on ecological aspects while improving the plan during the decision-making stage of the design's initial phases [15–17]. BIM entails showing a plan as an integration of "objects"—undefined and vague, product-specific or generic, void-space, directional, or solid shapes (such as a room's shape), which carry their relations, attributes, and geometry. BIM deals with more than just geometry. It also deals with geographic data, light analysis, spatial relationships, and properties and quantities of construction components (for instance, producers' details). BIM simulation also defines objects parametrically; so, objects are described as relationships and parameters to other objects so that dependent objects may always automatically adjust if a similar object is altered [18]. As BIM permits for multidisciplinary data to be superposed by sole modeling, it provides an opportunity for ecological evaluation methods, like life cycle analysis (LCA), to be included over the course of the planned procedure [19]. Thus, BIM increases the utilization of examining plan resolutions to develop ecological efficiency [20]. BIM equipment is too beneficial for designers and owners to choose the most proper energy-efficient plan for a design in the planning stage. At this stage, investing a smaller part of the project budget is ideal, before proceeding with more intense investing [21]. Given the diverse materials and designs before building a construction, the endeavor is to decrease ecological effects and diminish the life cycle expense of a construction [22]. BIM analysis also enables comprehensive definitions for the correct analysis of the energy consumption of the materials of a construction, so BIM must be used for defining the most useful plan in terms of performance in pre-construction steps. There are several construction materials providing better performance for reducing energy consumption, and they can be defined with the sole comparison table such as the undersurface used for this case study construction [23,24].

In this study, this popular tool is concentrated on external walls of the structures to compute the cooling and heating loads. Most constructions don't utilize bioclimatic configuration characteristics as utilized in conventional buildings with natural stone walls to diminish energy expenditure. As a conclusion, the evaluation of the climatic reactive of these housings depends on a quantitative and qualitative scrutiny. Conventional buildings are not equipped with cooling or/and heating tracts. They supply minimum comfort, especially in the summer months. In this case, the assessment of the efficiency will depend on thermal comfort levels. So the scrutiny will also depend on the evaluation of cooling and heating loads. A proper configuration designed using a regional structure can provide thermal comfort and integrate climate in both winter and summer months. The main aims of this study are defined as follows:

- Reaching a satisfying stabilization between environmental and economic liabilities with energy efficiency.
- Organizing interior design with the inclusion of energy efficiency to provide the best utilization of natural sources and minimal ecological effect.
- To research all the fields of energy utilization in all parts of the building and install structure for minimum degradation with characteristic goals.
- Looking for ways to add ecological characteristics into the next decision-making at all stages. BIM software program can help in raising maintainability in the building sector. It provides the analysis and integration of ecological topics in the plan over the building's life cycle [25].
- Trying to source materials from maintainable resources like any natural stone or other materials from the terrestrial elements that are traditional sources. Traditional resources have a natural and regional origin. They can be utilized as:
  - Supplementary materials, which are flooring materials (cork, clay, etc.) and materials used for painting and plastering.
  - Insulating materials, such as cork fiber, hemp fiber, and sheep's wool, etc.
  - Constructional materials, load-bearing structure, such as clay bricks, stone, wood, straw bales, rammed earth, etc. [26].

During the earlier investigations, researchers defined six steps of an optimal workflow from BIM to energy software [24]. These steps are: identifying the site of the construction materials, geometry, area types of construction, assigning areas as thermal regions, assigning lighting loads, occupants, appliances to be used in the area, identifying the heating, ventilation, and air conditioning (HVAC) tract and its elements in detail, and obtaining an energy analysis.

The researcher Eastman and his coworkers described BIM as the parametric modeling process and technique to communicate, analyze, and produce a digital structural modeling [27]. Given that BIM concentrates on using data associated with the entire life cycle of a structure plan, working together is too beneficial for an integrated and collaborative study among overall partners by an integrated project delivery strategy. Bazjanac commented on the significance of BIM simulation to be working together, which helps us to understand data better [28]. BEM simulation forecast of energy expenditure of constructions depends on mass flows and simulations of energy, contributing to decision-making in the early steps of a project [29]. This analysis results in thermodynamic principles, complex assumptions, and equations. They depend on elemental entry characteristics, such as the construction geometry, and complicated entry characteristics, such as the data associated with ventilation, air conditioning systems, and heating and ingredients, operation conditions, and weather [30]. Several researchers focus on the BEM–BIM design process in their articles [31]–[34]. Many other researchers especially define the BEM–BIM management and process throughout the planned procedure. Aida [35] and Motawa and Almarshad [36] suggest a determination of the plan procedure for BEM–BIM, including the essential determination perspectives, the data resources, and the data requirements for each of the planning phases. Likewise, Fan and Wong [37], Tuomas Laine [38], Lee [39] and Attia and their coworkers [40] define the required data for diverse plan phases to obtain a high-efficiency construction. How to integrate or link the BIM and BEM tools is the most challenging part of a BIM–BEM process; it was defined by Negendahl [41] and Toth Janssen [42] in their studies. Pezeshki and his coworkers, for example, represented a precious research on green-based BIM researches from 2015 to 2018 with a focus on the BIM database utilized in building energy modeling. Gao and his coworkers performed research on BIM-sourced building energy modeling for the improvement of energy performance in construction plans. Memari and Kamel researched the resolutions and challenges of working with BEM and BIM procedures. Sanhudo and his coworkers conducted a research about the technological ability of the BIM for strengthening energy performance. In this study, BIM simulation program is used in energy efficiency design of traditional buildings. In order to utilize a BIM-source energy performance ecologically, the whole procedure must be able to be actualized in an efficient and fully automated manner after startup. The utilization of BIM ensures many advantages linked with a common data modeling, which can vary from cost and time savings to a diminished risk of plan errors and collisions [43]. It can also be utilized during a construction life cycle by the owner, plan team, facility manager, and contractor as part of the refurbishment, maintenance, and ongoing monitoring programs [44]. As parts of the BIM procedure, BIM development simulations are utilized to design a primary construction plan conception. Then this primary plan is iteratively advanced in the next plan's steps depending on a simulation of its efficiency utilizing "BEM" implementations. The BEM implementations can also be utilized for aims such as energy consumption level control, according to construction arrangements, or assistance for developing an authentication procedure [45,46].

This study suggests an extended comparative BIM-based assessment model that highlights two purposes: (1) BIM-based economic assessment methodology, which aims to accomplish a less energy cost for the construction and (2) a comparative energy expenditure assessment methodology utilizing certain roof and floor design of the project after deciding on the most optimal natural stone wall material in the early plan step. Depending on the suggested status, the analysis was performed for the casework of an authentic restaurant situated in the Harput region in Elazig. The BIM-based methodology assesses the building's energy performance according to the energy costs and energy efficiency.

5.2 METHODOLOGY

The current application of BEM in the energy field contains an expert planning, a construction, and an energy expert. This energy expert manually uses a BEM simulation software and adds information on absent materials such as U-values (or heat transfer coefficients) [47]. Automating information change can improves this procedure, resulting in time savings and a decrease in working model mistakes by providing a reproducible and synchronized modeling. Frequently changing construction information relies on switching between modeling views of a construction so that information can be simulated by diverse fields. In the energy fields, an "architectonic view" is frequently changed into a "heat view" [47,48] and such conversion is essential for diverse causes. For instance, simulation tools frequently stand alone with diverse ways to present a modeling and issues required to be matched between one and the other.

BIM is a beneficial tool that can be used as a basis from which information could be removed to diverse fields [49], such as energy or daylighting [50,51]. BEM is also provides an opportunity to hold information as central source and this means staying away from information repetition and redundancy. Planners could develop their construction iteratively by modifying, for instance, energy requirement of a construction in a certain time. Thus, the investigation into the working together of Building Energy Modeling and Building Performance Simulation equipment solves many unsolved problems until now [52–54]. This shows us that information is frequently converted utilizing the Industry Foundation Class data model and the gbXML plan. The advantage of using gbXML plans is similar to using many energy equipment and BIM [55].

The workflow framework for energy efficiency simulation of the authentic restaurant building walls is given in Figure 5.1.

5.2.1 THE PROJECTING OF THE AUTHENTIC RESTAURANT

For a long time, architects and builders have been utilizing many features of natural stone to decorate the exterior and interior of structures. This natural building material represents individuality by its appearance and quality, thus original shells can be designed. In addition, construction with natural stone is specifically sustainable. The natural material has a wide scale of applications in modern architectural designs and also in the conservation and renovations of monumental structures. Natural stones highly vary in terms of their technical features such as frost resistance, heat storage capacity, compression strength, and water absorption. A natural stone's appearance doesn't reveal anything about its characteristics: a high-quality and affordable variants have the same outward look but have completely diverse properties. Soft natural stones are more proper for the interior, while harder and insulating natural stones are more proper for the exterior. In this study, three diverse natural stones are used

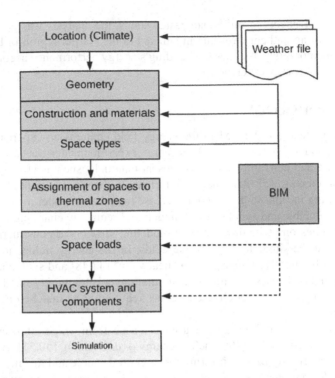

FIGURE 5.1 The workflow framework for energy efficiency simulation of the authentic restaurant building walls [56].

for the exterior wall design of the authentic restaurant. All of the stones have certain and durable insulating properties. This building is designed as a modern indoor and an authentic external wall.

The project of the authentic restaurant building is displayed in Figure 5.2.

The rendering of the authentic restaurant building are obtained by Lumion 3D rendering software and shown in Figure 5.3.

5.2.2 BUILDING COMPONENTS USED IN THE ANALYSIS OF THE AUTHENTIC RESTAURANT

The roof and ground layers used in designing the authentic restaurant building are shown in Figure 5.4. The wall structures used in the analysis of the authentic restaurant building are displayed in Figure 5.5.

The geological composition is significant for the evaluation of the natural stone. Silica and calcium are the two simple minerals available at different rates in all natural stones. Sometimes other different materials can be also found in the contents of a natural stone. The choice of natural stone affects its performance, characteristic properties, and mineral composition.

The required amount of energy of cooling and heating is forecasted by BIM, is similar to that for the required loads. It is evident that utilizing any of the proposed

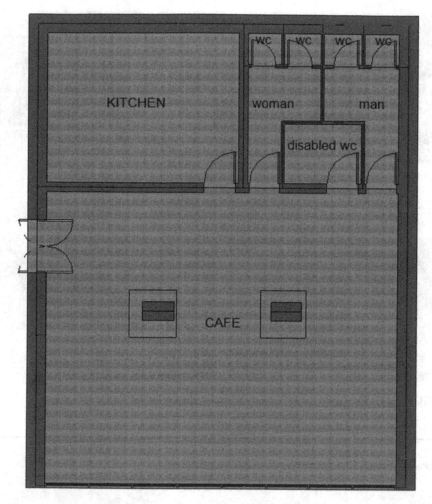

FIGURE 5.2 The project of the authentic restaurant building.

wall combinations (U-value) will lead to a degradation or increase in both cooling and heating energy needs throughout the year with certain ratio compared with that at the reference status [9]. This explicitly shows that developing the thermal capacity of external walls and using of external wall type suggested will be favorable and will give rise to substantial developments in the structure's thermal efficiency in the Harput region.

The thermal properties of wall materials (natural stones) used in the analysis are presented in Table 5.1.

The thermal conductivity coefficient properties of some wall materials defined at TSE 825 are given in Table 5.2, which presents a comparison of some wall materials defined at TSE 825 of natural stones analyzed in this study.

FIGURE 5.3   The rendering of the authentic restaurant building.

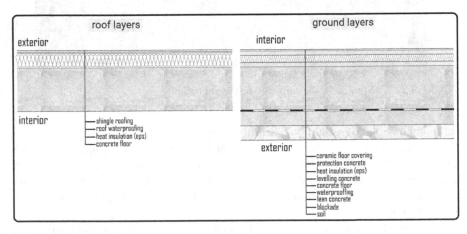

FIGURE 5.4   The roof and ground layers used in projecting the authentic restaurant building.

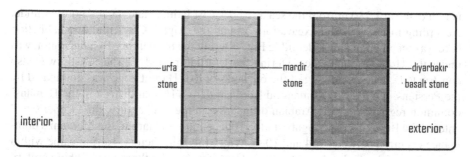

FIGURE 5.5 The wall structures used in the analysis of the authentic restaurant building.

TABLE 5.1
The Thermal Properties of Wall Materials (Natural Stones) Used in the Analysis

	Heat transfer coefficient (U) W/(m²K)	Thermal resistance (R) (m²K)/W	Thermal mass (B_{th}) kJ/K
Urfa stone	2.3667	0.4225	149.13
Mardin stone	1.3308	0.7514	86.98
Diyarbakır basalt stone	2.2750	0.4396	4.23

TABLE 5.2
The Thermal Conductivity Coefficient Properties of Some Wall Materials in the Literature [9]

		Values in literature	
Material	Density (gr/cm³)	$T_{avr.}$ (°C)	k (W/mK)
Concrete	2.272	20	1.512
Gas concrete	0.800	20	0.383
Brick	1.900	20	0.972
Ceramic	2.000	20	0.988

5.3 THE HISTORICAL FRAMEWORK, CLIMATIC FACTORS, AND LOCATION OF THE HARPUT REGION

5.3.1 THE HISTORICAL FRAMEWORK AND LOCATION OF THE HARPUT REGION

The authentic restaurant is planned to be built in the Harput region in Eastern Anatolia Region in Turkey. Harput region, also known as the "Upper City," is located within the borders of the Elazig province as a historical and touristic region. Harput is found above 1,280 m from the sea at 39°15′ east latitude and 48°43′ north latitude. According to research and excavations around the "Upper City", the first habitation here was recorded in the "Paleolithic Era". During the 8th century BC, the region was under the Urartian dominance. During the Urartian period, Harput Castle was also constructed. At the beginning of the 6th century BC, Harput zone was dominated by the Persians. In AD 379, Romans added the Valens's Harput district in the Romans' dominion region. After the Arabian occupation in the 7th century, the "Upper City" entered the Büveyhs' sovereignty till the 10th century. In Harput, the Mervani's dominance occurred between 954 and 1085. With Malazgirt victory, Harput came within the borders of the Turks. After this time, Harput was under the rule of Çubukoğulları, Artukids, Seljuks, Mongols, Ilkhanians, Dulkadiroğulları, and Akkoyunlu. In the post-republic period, Harput took the name "Elazig" [57].

This study deals with the design of the authentic restaurant in the Harput region in Elazig, in order to increase energy efficiency, to support ecological protection, and to sustain the authenticity of the ambience. With this aim, considering the historical texture of Harput region mentioned above, the optimal energy performance of the authentic restaurant designed with three different traditional natural stones was determined.

5.3.2 THE CLIMATIC FACTORS AFFECTING THE ANALYSIS OF THE HARPUT REGION

The regulations on day-degree regions are assessed according to TS 825 standard "Thermal Insulation Requirements for Buildings." The Turkish Standards Institution and Turkish State Meteorological Service grouped Turkish climate zones as "thermal insulation zones" by utilizing a degree-day methodology, which was improved by the Turkish State Meteorological Service. According to this classification, Turkey is separated into four insulation zones [9,58].

The temperatures over 10°C, which is obtained from 236 weather stations from 1981 to 2001, has been computed as below:

Efficient total temperature (degree-days) = (number of days in the month) *
(monthly mean temperature – 10).

In this formulation, degree-days for all provinces and some towns are given on the basis of the monthly basis average temperatures. By means of values defined, Turkey's 81 provinces are clearly separated into four degree-day regions to enable heating energy load computations in TS 825. By this classification, determining consumption values and insulation requirements for TSE 825 have been feasible. In this way, heating energy needs are determined, construction design characteristics are recommended, and climate information is given for all regions. The location of Elazig city, according to degree-day regions over Turkey, is given in Figure 5.6 [9].

The solar map (a) and wind map (b) of Elazig city are displayed in Figure 5.7 [58,59].

The mean low and high temperature variation of Elazig city is displayed in Figure 5.8(a). In this city, the winters are very cold, dry, and partly cloudy, and the

FIGURE 5.6 (a) The location of Elazig city, according to degree-day regions over Turkey [9]. (b) Harput region architectural constructions.

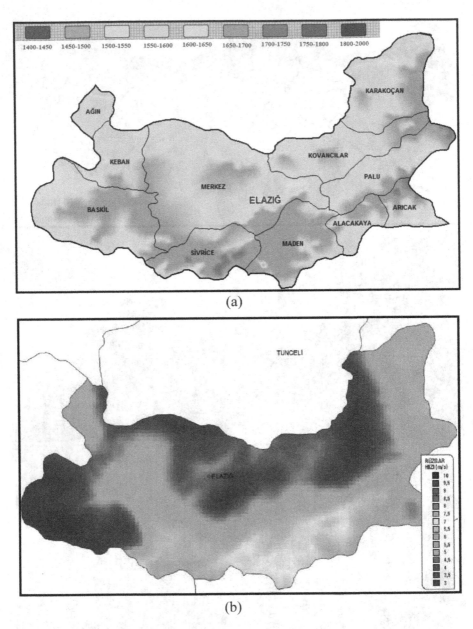

| 1400-1450 | 1450-1500 | 1500-1550 | 1550-1600 | 1600-1650 | 1650-1700 | 1700-1750 | 1750-1800 | 1800-2000 |

(a)

(b)

FIGURE 5.7   (a) The solar map of Elazig city [54]. (b) The wind map of Elazig city [59].

summers are hot, arid, and mostly clear. Throughout the year, temperatures usually range from 94°F to 22°F. The hot months are from September to June. July 26 is the hottest day of the year. The cold months are from March to November. January 25 is the coldest day of the year (Figure 5.8(a)). The daily mean low and high warmth are with 90th to 10th and 75th to 25th percentage tapes. Fine dotted drawings correspond

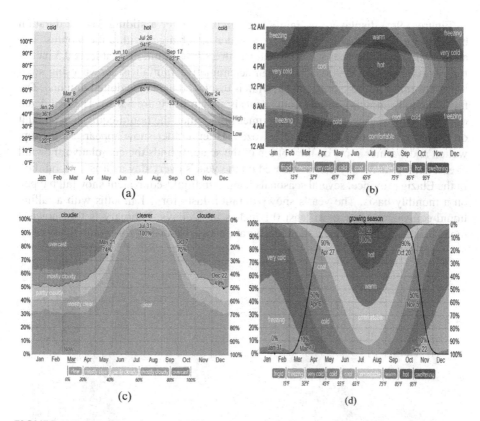

FIGURE 5.8 (a) Mean low and high temperature variation. (b) Mean temperature on an hourly basis. (c) Cloud cover categories. (d) Period of time in different growth seasons and temperature bands.

to the mean sensed warmth. In Elazig city, the mean hourly temperature is shown in Figure 5.8(b). The mean hourly warmth is shown using the colored encoded tapes. The shaded coverings display twilight and night. This figure reveals an annual compact qualification. The change in color shows the mean temperature of that day and hour. The vertical axis shows the day's hours, and the horizontal axis shows the year's days. Cloud cover categories of Elazig city are given in Figure 5.8(c). The percent period of time in each of the cloud covering bands is classified via the covered sky percent by means of the clouds. In Elazig city, the mean cloudiness percentage faces the utmost seasonal change throughout the year. The clearer part and the cloudier part of the year last for 4.5 and 7.5 months, respectively. Growth seasons and temperature bands in different time periods are shown in Figure 5.8(d). The percent of opportunity that a certain day is within the season is shown using the black line. The season's definitions vary worldwide. The figure describes the longest continuous nonfreezing temperature period (almost 32°F) per annum. (The calendar year of the Northern Hemisphere or the Southern Hemisphere from July 1 to June 30.). In Elazig city, the season usually lasts for seven months [60].

Figure 5.9(a) displayed the rainfall accumulated over a sliding 31-day duration based around every day of the year to consider the changes within the months. The annual rainy duration lasts for 8.5 months with a sliding rainfall of at least 0.5 inches for 31st day (Figure 5.9(a)). The mean accumulative rainfall is shown using solid lines. The mean liquid-equivalent snowfall is displayed using the dotted thin line. The mean monthly snowfall (liquid-equivalent) is shown in Figure 5.8(b). The real depth of novel snowfall is usually five to ten times higher than the liquid-equivalent amount on the assumption that the ground is frozen. Drier, colder snow appears on the top, wetter, and warmer snow on the bottom of the scale. It finds the cumulative snowfall over a sliding of 31-day duration based on the year's every day as it is with rainfall. In the Elazig province, several seasonal changes in liquid-equivalent snowfall happen on a monthly basis. The year's snowy duration lasts for 4.1 months with a falling liquid-equivalent snow of at least 0.1 inches in 31 days. February is the snowiest month of the year. The year's snowless duration lasts for over a 7.9-month time period

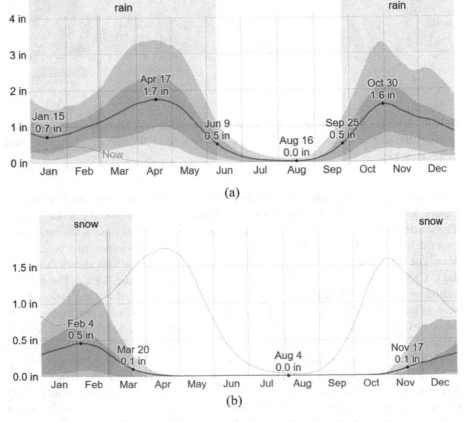

FIGURE 5.9 (a) Mean rainfall on a monthly basis. (b) Mean snowfall (liquid-equivalent) on a monthly basis.

(Figure 5.9(b)). The mean cumulative liquid-equivalent snowfall is shown using solid lines. The mean rainfall is displayed using the dotted thin line [60].

Figure 5.10(a) displays the wide area hourly mean wind vector value (direction and speed) at 10 m height above the ground. The wind occurring in an area is greatly dependent on regional topography and other variables. In Elazig province, the mean hourly wind velocity occurs with a mild seasonal change over a year period. The year's windswept part lasts for over 2.9 months with mean wind velocities of more than 10.29 km (6.4 miles) in the hour. The year's calmer period lasts for 9.1 months. With 75th to 25th and 90th to 10th percentage bands, the mean average hourly wind velocities are shown using dark-gray line. Figure 5.10(b) shows wind directions throughout a year in Elazig province. The predominant mean wind direction changes hourly over the course of the year. In Elazig province, respectively, the most often

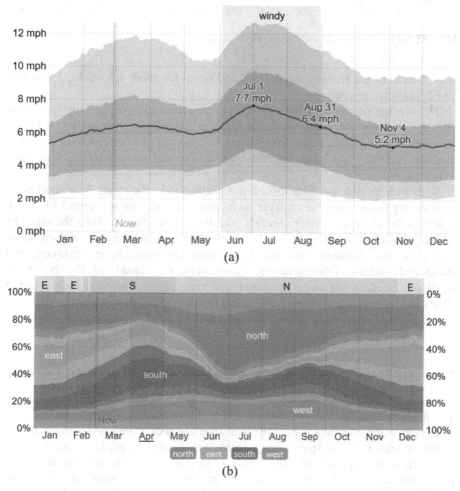

FIGURE 5.10 (a) Average wind speed. (b) Wind direction.

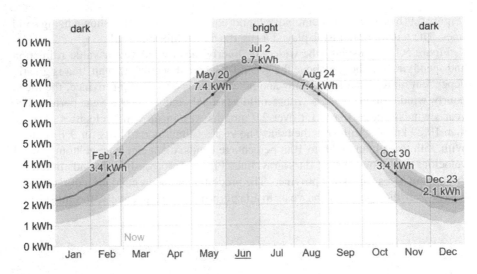

FIGURE 5.11   Mean daily event short-wave sun power.

wind directions are from the north, east, and south for about 1.3 weeks, 2.7 weeks, and 2.6 months, respectively.

The average percent of wind directions comes from each of the four main wind directions, except the hours when the mean wind velocity is less than 1.0 mph. The lightly tinted fields at the frontiers are the amount of hours spent in the intermediate directions (southeast, northeast, northwest, and southwest) that have been indicated [60].

With 90th to 10th and 75th to 25th percentile bands, the mean daily short-wave solar energy that reaches the earth per m² (orange line) is given in Figure 5.11, which describes the short-wave solar energy's overall daily incident reaching the ground's surface throughout a wide field. In addition to taking into account seasonal variations fully in daytime duration, absorption by the means of atmospheric constituents and clouds and elevation of the sun above the horizon are also considered. The short-wave irradiation contains ultraviolet irradiation and visible light. The mean daily event of short-wave radiation and excessive seasonal change occurs throughout the year. The year's brightest duration lasts for 3.1 months and the year's darkest duration lasts for over 3.6 months [60].

5.4   THE ENERGY EFFICIENCY ANALYSIS WITH BUILDING ENERGY MODELING OF THE AUTHENTIC RESTAURANT BUILDING

The methodology of utilizing BIM for the building modeling process has become an attractive and prevalent in both the industrialized society and in the research field in recent years [61–63]. In this study, BEM is particularly a perfect example of the early plan step, when suitable approaches and the most effective cost for the energy performance plan can be integrated with the entire construction plan procedure. BIM can begin early even in the pre-plan stage and even in the schematic plan stage in the notion plan step. This will guide the plan to continue to the detailed plan steps so as to take appropriate decisions. The primary outcomes of BEM are shown below [64–66].

- Computer analysis of the entire construction, such as energy cost, energy use, and energy consumption
- Improved energy performance level
- Achievement of energy review and simulation for compliance requirements with LEED projects and implied energy codes
- Compliance code analysis
- Construction system's optimization for energy expenditure to save on productivity and ensure user comfort
- Construction cost effectiveness and design optimization
- Construction cost effectiveness and operational optimization
- Operation cost
- Lifecycle costing analysis and energy modeling analysis

In this study, energy consumption, energy cost, and the energy use of an authentic restaurant designed in the Harput region of Turkey are investigated by BEM.

5.4.1 THE EVALUATION OF ENERGY CONSUMPTION OF THE AUTHENTIC RESTAURANT'S BUILDING WALLS

The thermal features of wall building materials have an important impact on evaluating the thermal quality of constructions. Hence, choosing the suitable wall components is important for forming energy-effective constructions that consumer lower amount of energy to sustain comfort in closed areas [67]. It is apparent that this aim of the construction plan is conducive to form a more maintainable built environment, with minimal inverse impact on the ecology [68]. It has significance in the Harput region and worldwide for immediate energy-saving requirements and also build energy into plans in the future. To analyze the LCC (life cycle cost) and the operation costs, the efficiency of the construction and construction's components must be estimated. Toward this aim, BIM with its combined energy assessment equipment was utilized to estimate the energy requirements of the construction. Upon creating "thermal areas" in Revit, which reflect different building sectors, the BIM utilizes software to collect internal temperatures using a downloaded climate file [69,70]. These hourly temperature data vary depending on the assembly of buildings, and any structure that would impact solar benefits such as overheating. These temperatures are viewed better as a long-term average that includes weeks or months, rather than an hourly average. They are also often used to measure the impact of a structure of a building on another. These temperatures are derived from a climate study that uses average temperatures for over 30 years.

The research is carried out by transforming the overall shape and layout of buildings into a "computational network" capable of capturing essentially all the primary paths and processes of heat transfer throughout the building. It evaluates the results of the possible energy performance of building models with existing external wall types, depending on the materials used to create the building model. In this analysis, the indoor and outdoor temperatures are used as 22°C and −12°C, respectively.

In the evaluation of the energy performance of the authentic restaurant building in Harput region of Elazig city (Turkey), different natural stones used as wall material were compared by using a project designed with the same roof and floor materials. The annual energy performance values (MJ) of the authentic restaurant building walls are displayed in Figure 5.12.

Monthly energy performance values (MJ) of the authentic restaurant building walls are shown in Figure 5.13.

Among the natural stone wall materials, the highest energy consumption and energy cost values through external walls designed with the Urfa stone are obtained as 466,071 MJ and 9,378 T per m², respectively. The T unit here is the abbreviation of the Turkish lira. With the Urfa stone used as external wall material, the maximum and minimum total energy consumptions over the course of a year are observed in January as 72,944 MJ and in September as 18,527 MJ. Among energy-consuming elements of a building, the highest energy consumption was found in space heating as 232,057 MJ in the total consumption. The lowest energy consumption was determined with pumps aux (2,654 MJ). By using Mardin stone, the minimum energy consumption values were also determined (452,796 MJ). Over the course of a year, the highest and lowest total energy consumptions are defined in January as 70,373 MJ and in September as 19,508 MJ with Mardin stone used as external wall material.

Among the external walls, the maximum energy consumption areas are determined in the order as follows: space heating (232,057 MJ), hot water (113,818 MJ), miscellaneous equipment (45,954 MJ), area lights (35,735 MJ), space cooling (19,065 MJ), vent fans (19,040 MJ), and pumps aux (2,654 MJ). By using the Diyarbakır stone, energy consumption is observed as 452,960 MJ for a period of one year. The highest and lowest total energy consumptions are annually obtained in January (70,353 MJ) and September (19,508 MJ), respectively. Among energy-consuming elements of a building, the highest energy consumption was found in space heating with a value of 694 GJ. The lowest energy-consuming determined with pumps aux was 2,654 MJ. In this analysis, it was found that Mardin stone wall gave us the most effective results among all the tested wall types. The worst energy performance was obtained from the Urfa stone wall. During the building analysis, the highest energy-consuming element found was space heating.

The highest values of natural gas and electrical energy consumption determined for the periods of December–January–February and June–July–August have been determined among all external wall styles. The highest values of total energy consumption are observed for the periods of December–January–February.

5.4.2 The Evaluation of the Energy Cost of the Authentic Restaurant Building Walls

BEM is a very effective tool in diminishing, predicting energy costs, sizing, optimizing of energy systems, and ensuring options for energy resolutions. The information on annual cost and energy consumption will provide data on energy cost assessments for construction and early design decisions. Building energy expense estimates are

Annual energy consumption (MJ)	Natural stone type		
	Urfa Stone	Mardin Stone	Diyarbakır Basalt Stone
Area lights	35,735	35,735	35,735
Miscellaneous equipment	45,954	45,954	45,954
Space cooling	17,401	17,212	19,065
Vent fans	18,452	17,496	19,040
Pumps aux	2,654	2,654	2,654
Space heating	232,057	219,927	216,694
Hot water	113,818	113,818	113,818
Total	466,071	452,796	452,960

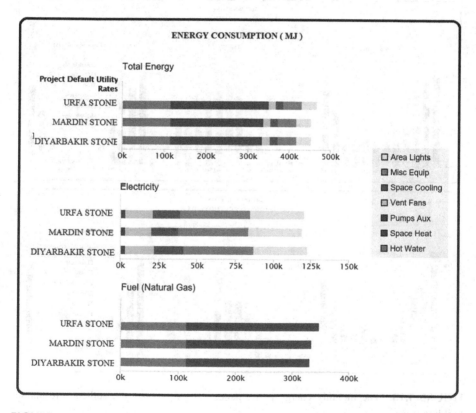

FIGURE 5.12 The annual energy performance values (MJ) of the authentic restaurant building walls

applied to the lifecycle cost, providing a complete picture for the decision-making of investors and owners. Using alternative systems or materials can increase the energy cost of the building and result in the opposite outcome. Many building energy cost calculations and early compliance decisions could have been made using the annual

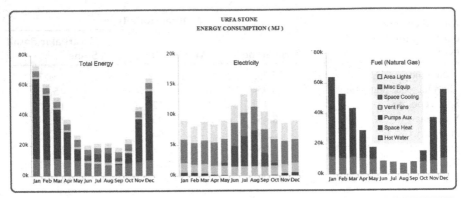

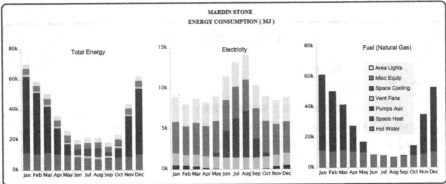

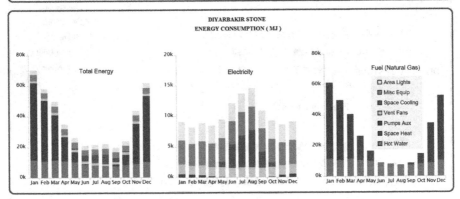

FIGURE 5.13   The monthly energy performance values (MJ) of the authentic restaurant building walls

energy cost and usage data. The costs are calculated using the average, the country and nationwide utility rates, or the custom rates that may have contributed to the energy expenses. Comparing costs can be daunting, as energy is billed in mega joule (MJ) for thermal natural gas. The current annual energy cost T per m² is reported by the energy costs. The modification of the input will instantly change this value. The energy costs are obtained by using mean utility ratios through simulation software across the region, nation, or statewide [35].

During the design phase, we must implement a strategy to effectively minimize the cost of energy usage of a building. Besides this, if we make a smart design, a 50–75 percent reduction in energy consumption can be achieved compared to buildings completed in the 2000s. Thus, energy-based costs will also decline [36–38]. The energy costs were determined as a linear ratio with energy needs for the analyzed external wall forms.

The annual energy cost values (T per m^2) of the authentic restaurant building walls are given in Figure 5.14. As shown in Figure 5.14, even if electricity prices are considered in the range, the natural gas pricing is about three to four times lower than the electricity pricing.

The monthly energy cost values (T per m^2) of the authentic restaurant building walls are shown in Figure 5.15.

By examining the types of external walls, this study was designed to identify the groups of power limits within a development stage model and to quantitatively evaluate how these groups could influence the projected energy usage. Figure 5.14 has shown the remarkable shift in building energy costs (T per m^2) with natural stone external wall forms comprising higher performance external wall materials, thereby demonstrating its important role in improving the energy performance of the right building shell.

The energy cost varies between 9,378 T per m^2 and 9,160 T per m^2. The energy cost reduced with the decrease in energy consumption in all the natural stone external wall types. The highest total energy cost value was obtained with Urfa stone external wall as 9,378 T per m^2. The lowest value of total energy cost, 9,160 T per m^2, was obtained using the Mardin stone as external wall material. At the authentic restaurant building walls, the total energy cost values of the external wall with Diyarbakır and Urfa stones increased by 1.98 and 0.99 percent, respectively, compared to those of the external wall with Mardin stone. The external wall with Urfa stone has given the highest energy cost in January, which was 0.90 percent higher than in December, 0.84 percent higher than in February, and 0.76 percent higher than in March.

5.4.3 EVALUATION OF ENERGY USAGE INTENSITY OF THE AUTHENTIC RESTAURANT'S BUILDING WALLS

It is difficult to compare the energy usage between structures or buildings in the absence of a norm or a benchmark. The simple calculation of the amount of energy used per specified period is not taken into account in terms of the scale, the form, the configuration, or the usage of the building. The practice of an index of energy usage intensity provides the means to equalize the way energy usage can be measured between different types of buildings and to determine the means of diminishing overall energy consumption. Due to the differences in the cooling and heating costs between diverse regions of the country, the climate can have an important impact on energy usage intensity. Therefore, we can divide energy usage intensity values into regions to provide a more accurate comparison of selected buildings, or we can use

Annual energy cost (T)	Natural stone type		
	Urfa Stone	Mardin Stone	Diyarbakır Basalt Stone
Area lights	1,377	1,377	1,377
Miscellaneous equipment	1,777	1,777	1,777
Space cooling	671	664	735
Vent fans	718	675	730
Pumps aux	102	102	103
Space heat	3,177	3,009	2,965
Hot water	1,556	1,556	1,556
Total	9,378	9,160	9,243

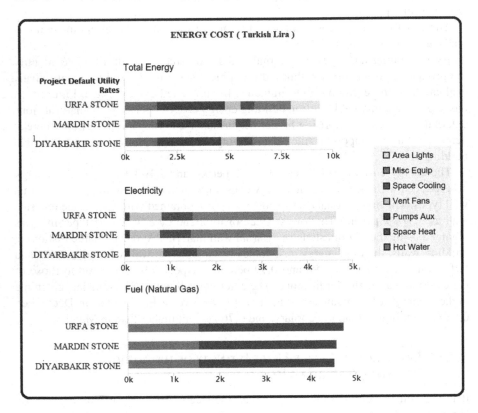

FIGURE 5.14 The annual energy cost values (T per m²) of the authentic restaurant building walls

"normalized-weather" values to adjust the energy usage intensity to be equivalent to a building in a diverse climate. As a result, energy usage intensity can also significantly vary depending on the type of the building. The energy usage intensity expresses itself as electricity consumed per square foot per annum. It is calculated by dividing the building's total energy consumption in one year (measured in kBtu or MJ) by the building's total gross floor area.

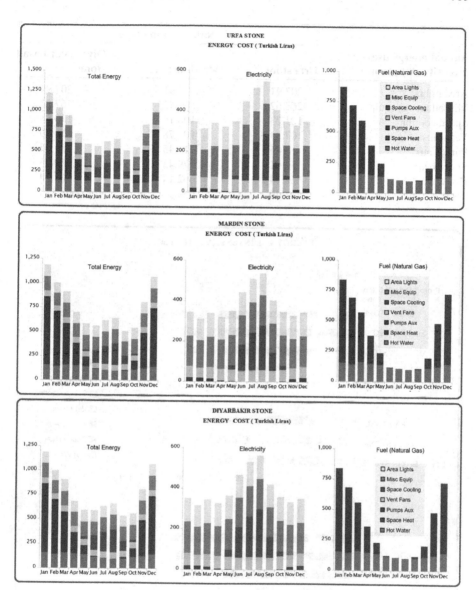

FIGURE 5.15 The monthly energy cost values (T per m²) of the authentic restaurant building walls.

Comparing the energy requirements of different building sizes gives a clear picture of energy consumption. Therefore, it can show us which method is the best to apply. Energy usage intensity is a particularly useful method for setting energy usage goals and benchmarks. It usually varies depending on the environment, the construction design, and the size of the building. Energy usage intensity is the sum of the combined fuel and energy per area of the project per year (square feet in IP units or square meters in SI units). The annual energy usage intensity values (MJ per m²) of the authentic restaurant building walls are given in Figure 5.16. This only takes into

Annual energy use intensity (MJ per m²)	Natural stone type		
	Urfa stone	Mardin stone	Diyarbakır basalt stone
Area lights	207.91	207.91	207.91
Miscellaneous equip	267.62	267.32	267.32
Space cooling	101.23	100.14	110.9
Vent fans	107.34	101.78	110.79
Pumps aux	15.45	15.45	15.45
Space heat	1,350.01	1,279.43	1,260.63
Hot water	662.14	662.14	662.14
Total	2,711.40	2,634.17	2,635.14

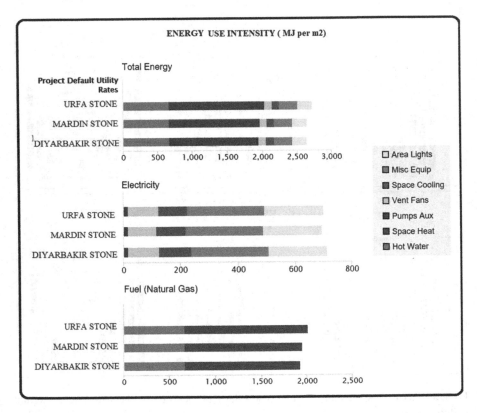

FIGURE 5.16 The annual energy use intensity values (MJ per m²) of the authentic restaurant building walls.

account the amount of fuel and electricity used at the site ("location" or "secondary" energy). We do not know the fuel used to generate electricity or heat energy. Most energy usage intensity tabulations and building codes aim to capture the full effect of providing energy to a building. This is done by defining the terms "source" or "key" energy and combining the fuel used to generate power on-site or at a far-off power station. For example, 1 kWh of site electricity from the solar panel on the building's roof is equal to 1 kWh of "source" energy, as the resource is the solar panel itself. The energy location is the most useful measurement of energy usage, providing visual or thermal comfort while measuring. However, while measuring total energy usage to determine environmental effects, the source energy is the most accurate measurement tool.

Among the authentic restaurant external wall types, the lowest energy usage intensity determined as 2,634.17 MJ per m^2 with Mardin stone. The highest energy usage intensity determined as 2,711.4 MJ per m^2 with Urfa stone. The maximum and minimum energy usage intensities were obtained from space heating and pumps aux, respectively.

The area lights, miscellaneous equipment, space cooling, ventilating fans, pumps auxiliaries, space heating, and hot water values are determined for Urfa stones as 207.91, 267.32, 101.23, 107.34, 15.45, 1,350.01, 662.14 MJ, for Diyarbakır stone as 207.91, 267.32, 110.9, 110.79, 15.45, 1,260.63, 662.14 MJ, and for Mardin stones as 9,207.91, 267.32, 100.14, 101.78, 15.45, 1,279.43, 662.14 MJ, respectively. The energy productivity of the authentic designs is important. Lately, this type of authentic new buildings has been built in order not to spoil the historical appearance of many historical cities in Turkey. These buildings are supported with energy to ensure specific cooling conditions. The total energy consumption of these buildings, which were built in large numbers, should be controlled even if they were within a small area. For this purpose, the most logical solution is to create more efficient buildings, in terms of energy performance, by choosing the right building materials while designing the building. It is possible to contribute to the reduction of energy consumption throughout the country with these buildings, which are designed in the most suitable way for the climatic conditions of the region.

5.5 CONCLUSIONS

For determining the energy efficiency of a construction and to assess the architectural and construction systems' plan, BIM is employed. Compared to the rule of thumb, which consists of extrapolation methodologies utilized by planners and manual computations, BEM can ensure more reliable conclusions and better energy-efficient plan into the general construction plan procedure. BIM is considered as a numerical model procedure of managing and generating the stable construction data during the construction's lifecycle. In addition to the captured data about the construction, other decision-making tools can be employed to achieve the maintainability targets and to optimize the construction efficiency. BIM-based BEM is a common tool used in the BIM industry, which has the potential to optimize the cost, time of BIM, to assess diverse plan options rapidly, to develop BIM correctly, and to obtain better construction energy efficiency; see Table 5.3.

Unlike human-made products, such as concrete and cast stone, natural stone extracted from the ground is used to build the external wall. It is an entirely natural product, processed utilizing lower energy while converting into a building material

TABLE 5.3
The Evaluation Results of Wall Materials (Natural Stones) Used in the Analysis

	Natural stone types		
	Urfa stone	Mardin stone	Diyarbakır Basalt stone
Annual energy consumption (MJ)	466 071	452 796	452 960
Annual energy cost (T)	9378	9160	9243
Annual energy use intensity (MJ per m²)	2711.40	2634.17	2635.14

form. The production process of concrete construction materials contains high energy conversion in fuel-burning furnaces, which pumps huge amounts of carbon dioxide to the atmosphere. Before polluting, resins are mixed with chemicals to bind it all together. The natural stone production does not require any of these practices.

For this reason, we have focused on the main energy cost and usage of natural stones for an authentic restaurant. In this study, we analyzed the various external wall materials' effects on the cost of primary energy and energy usage for designing and building the authentic restaurant. We compared three different natural stone materials and discussed how these materials could minimize the cost and the usage of primary energy production. To achieve this aim, the diverse designs of the authentic restaurant building walls are compared by using proper natural stone materials with respect to energy efficiency. In the Harput region in the Elazig city (Turkey), the authentic restaurant building walls are modeled by using Autodesk Revit Architecture (BIM tool). The best performance wall type is determined as the Mardin stone external wall. In the Mardin stone external wall, energy usage intensity, annual energy performance, and energy cost are defined as 2,634.17 MJ per m², 452,796 MJ, and 9,160 T per m², respectively. The worst performance wall type is determined as the Urfa stone external wall. In the Urfa stone external wall, energy usage intensity, annual energy performance, and energy cost are defined as 2,711.4 MJ per m², 466,071 MJ, and 9 578 T per m², respectively. Overall, it is possible that there is a link between natural stone density and thermal conductivity, especially if that reduction in density is caused by an increase in natural stone porosity. This study suggests that strategies for low-energy buildings should be combined with efficient and high-performance natural stone materials to mitigate climate change and for a sustainable built environment in authentic buildings. Our results display that the natural stone option used as the building envelope material has an important impact on the use of primary energy and energy costs of the alternatives studied. With this perspective, the BIM-based modeling could function as a significant decision tool about the direct-wall material impacts on the performance.

REFERENCES

[1] F. Balo. Feasibility study of "green" insulation materials including tall oil: Environmental, economical and thermal properties. *Energy Buildings* 86 (2015) 161–175.

[2]   F. Balo. Energy and economic analyses of insulated exterior walls for four different city in Turkey, Energy Education Science and Technology Part A: *Energy Sci. Res.*, 26(2) (2011) 175–188.

[3]   F Balo. Castor oil-based building materials reinforced with fly ash, clay, expanded perlite and pumice powder, Ceramics-Silikáty 55 (3) (2011) 280–293.

[4]   U. Bogenstätter, Prediction and optimization of life-cycle costs in early design, *Build. Res. Inf.* 28 (5–6) (2000) 376–386.

[5]   B. Kolarevic and A.M. Malkawi, *Performative Architecture: Beyond Instrumentality*, Spon Press, New York, 2005.

[6]   D.B. Crawley, J.W. Hand, M. Kummert, and B.T. Griffith, Contrasting the capabilities of building energy performance simulation programs, *Build. Environ.* 43 (4) (2008) 661–673.

[7]   A.M. Malkawi, Developments in environmental performance simulation. *Autom. Constr.* 13 (4) (2004) 437–445.

[8]   R. Oxman, Performance-based design: current practices and research issues. *Int. J. Archit. Comput.* 6 (1) (2008) 1–17.

[9]   TS 825. Thermal insulation requirements for buildings, 2008. TSE, Ankara, Turkey.

[10] N. D. Šekularac, D. M. Šumarac, J. L. Cˇikicˊ Tovarovicˊ, M. M. Cokic, and J. A. Ivanovicˊ-Šekularac. Re-use of historic buildings and energy refurbishment analysis via building performance simulation: A case study. *Therm. Sci.* 22 (2018) 2335–2354.

[11] V. Murgul. Features of energy efficient upgrade of historic buildings: Illustrated with the example of Saint-Petersburg. *J. Appl. Eng. Sci.* 12 (2014) 1–10.

[12] N. Sekularac, J. I. Sekularac, and J. C. Tovarovic, Application of stone as a roofing in the reconstruction and construction. *J. Civ. Eng. Archit.* 6 (2012) 919.

[13] K. Günçe, and D. Mısırlısoy. Assessment of adaptive reuse practices through user experiences: Traditional houses in the walled city of Nicosia. *Sustainability* 11 (2019) 540.

[14] B. Figen. Investigation of the usability of flax oil-based materials as building or insulation material in buildings, 11th National Installation Engineering Congress and Teskon + Sodex Fair, pp: 752–762, April 17–20, 2013, İzmir, Turkey.

[15] C. K. Anand and B. Amor. Recent developments, future challenges and new research directions in LCA of buildings: A critical review. *Renew Sustain Energy Rev.* 67 (2017) 408–416. https://doi.org/10.1016/j.rser.2016.09.058

[16] M. Röck, A. Hollberg, G. Habert, and A. Passer (2018a). LCA and BIM: Integrated assessment and visualization of building elements' embodied impacts for design guidance in early stages. In Procedia CIRP (Vol. 69, pp. 218–223). https://doi.org/10.1016/j.procir.2017.11.087.

[17] F. Shadram, T. D. Johansson, W. Lu, J. Schade, and T. Olofsson, An integrated BIM-based framework for minimizing embodied energy during building design. *Energy and Buildings*, 128 (2016) 592–604. https://doi.org/10.1016/j.enbuild.2016.07.007

[18] F. Leite, A. Akcamete, B. Akinci, G. Atasoy, and S. Kiziltas. Analysis of modeling effort and impact of different levels of detail in building information models. *Autom. Construct.* 20 (5) (2011) 601–609.

[19] F. Jalaei and A. Jrade, An automated BIM model to conceptually design, analyze, simulate, and assess sustainable building projects. *J. Construct. Eng.* (2014) 1–21. https://doi.org/10.1155/2014/672896

[20] C. Cavalliere, G. Habert, G. R. Dell'Osso, and A. Hollberg, Continuous BIM-based assessment of embodied environmental impacts throughout the design process. *J Cleaner Produc.* 211 (2019) 941–952. https://doi.org/10.1016/j.jclepro.2018.11.247

[21]   K. U. Ahn, Y. J. Kim, S. C. Park, I. Kim, and K. Lee, BIM interface for full vs. semi-automated building energy simulation, *J. Energy Buildings*, 68 (2013) 671–678.

[22]   E. Krygiel and B. Nies, *Green BIM: Successful Sustainable, Design with Building, Information Modeling*. Wiley, 2008.

[23]   S. Azhar. Building information modeling (BIM): trends, benefits, risks, and challenges for the AEC industry. *Leadersh Manage Eng.* 11 (3) (2011) 241–252.

[24]   G. Gourlis and I. Kovacic. Building information modelling for analysis of energy efficient industrial buildings—a case study. *Renew Sustain Energy Rev* 68 (2017) 953–963.

[25]   S. Maltese, L.C. Tagliabue, F.R. Cecconi, D. Pasini, M. Manfren, Ciribini, and L.C. Angelo. Sustainability assessment through green BIM for environmental, social and economic efficiency. *Procedia Eng.* 180 (2017) 520–530.

[26]   M. Spišáková and D. Mačková. The use potential of traditional building materials for the realization of structures by modern methods of construction. *SSP—J Civil Eng.* 10 (2) (2015) 127–138.

[27]   C. Eastman et al. *BIM Handbook: A Guide to Building Information Modeling for Owners, Managers, Designers, Engineers, and Contractors*. New York: John Wiley & Sons, 2008.

[28]   V. Bazjanac. Acquisition of building geometry in the simulation of energy performance. Proc. Intern. Conf., Rio de Janeiro, 1: 305–311. Lamberts, et al. (eds), ISBN 85-901939-2-6. 2001.

[29]   D. B. Crawley et al. EnergyPlus: creating a new- generation building energy simulationprogram. *Energy Buildings* 33 (4) (2001) 319–331.

[30]   V. Bazjanac and D. B. Crawley. Industry foundation classes and interoperable commercial software in support of design of energy-efficient buildings. Building Simulation '99, 6th International IBPSA Conference: 661–667, Udagawa and Hensen, ed., Kyoto, Japan. September 13–15, 1999.

[31]   S. Korkmaz et al. High-performance green building design process modeling and integrated use of visualization tools. *J. Archit. Eng.* 16 (1) (2010) 37–45.

[32]   T. Koppinen and A. Kiviniemi. *Requirements Management and Critical Decision Points*. JULKAISIJA – UTGIVARE, Finland, 2007. ISBN 978-951-38-7153-6. www.vtt.fi/inf/pdf/workingpapers/2007.

[33]   İ. Bahriye and H. Yaman. BIM and sustainable construction integration: An IFC-based model. *Megaron*. 10 (3) (2015) 440–448.

[34]   M.A. Zanni, R. Soetanto, and K. Ruikar, Towards a BIM-enabled sustainable building design process: Roles, responsibilities, and requirements, *Archit. Eng. Des. Manag.* 13 (2016) 101–129.

[35]   Farzaneh Aida Development of BIM-BEM Framework to Support the Design Process, Thesis, Montreal, December 11, 2019.

[36]   I. Motawa and A. Almarshad A knowledge-based BIM system for building mainten-ance. *Autom. Construc.t* 29 (2013) 173–182.

[37]   K.-d. Wong and Q. Fan. Building information modelling (BIM) for sustainable building design. *Facilities* 31 (3/4) (2013) 138–157.

[38]   T.J. Tuomas Laine. COBIM common BIM requirement, 2012.

[39   Y.G. Lee. Developing a design supporting system in the real-time manner for low energy building design based on BIM, in: Proceedings of the International Conference on Human–Computer Interaction, Springer, 2016.

[40]   S. Attia, E. Walter, and M. Andersen. Identifying and modeling the integrated design process of net zero energy buildings, in: Proceedings of the High Performance Buildings-Design and Evaluation Methodologies Conference, Brussels, 2013.

[41] M. Tsikos and K. Negendahl. Sustainable Design with Respect to LCA Using Parametric Design and BIM Tools. In World Sustainable Built Environment Conference 2017. Hong Kong.

[42] J.A. Toth, D. Crawley, S. Geissler, and G. Lindsey. Comparative assessment of environmental performance tools and the role of the Green Building Challenge. *Building Res. Inform.* 29(5) (2001) 324–335.

[43] L. H. Forbes and S. M. Ahmed. *Modern Construction: Lean Project Delivery and Integrated Practices.* CRC Press, Boca Raton, FL, 2011.

[44] D. Bryde, M. Broquetas, and J. M. Volm. The project benefits of Building Information Modelling (BIM). *Int. J. Proj. Manag.* 31 (7) (2013) 971–980.

[45] W. Keiholz, B. Ferries, F. Andrieux, and J. Noel, A simple, neutral building data model, In: *eWork and eBusiness in Architecture, Engineering and Construction.* London: Taylor & Francis Group, 2009, pp. 105–109.

[46] W. Feist. The Passive House Institute [Online]. Available: www.passiv.de/en/index.php2012

[47] V. Bazjanac and A. Kiviniemi. Reduction, simplification, translation and interpretation in the exchange of model data, In: Proceedings of the 24th CIB W78 Conference: Bringing ICT knowledge to work. University of Maribor 2007, pp. 163–168.

[48] C. Wilkins and A. Kiviniemi. Engineering-centric BIM. *ASHRAE J.* 50 (2008) 44–48.

[49] C. Eastman. *Building Product Models: Computer Environments Supporting Design and Construction,* Boca Raton, FL: CRC Press, 1999.

[50] E. Krygiel and B. Nies. *Green BIM.* Indianapolis, IN: Wiley Publishing, 2008.

[51] A. Osello, G. Cangialosi, D. Dalmasso, A. Di Paolo, M.L. Turco, P. Piumatti, and M. Vozzola. Architecture data and energy efficiency simulations: BIM and interoperability standards, Proceedings of Building Simulation 2011: 12th Conference of International Building Performance Simulation Association 2011, pp. 1521–1526 (Sydney), [Online]. Available: www.ibpsa.org/proceedings/BS2011/P_1702.pdf. (Accessed: January 5, 2020).

[52] H. J. Moon, M. S. Choi, S. K. Kim, and S. H. Ryu. Case studies for the evaluation of interoperability between a BIM based architectural model and building performance analysis programs, Proceedings of Building Simulation 2011: 12th Conference of International Building Performance Simulation Association 2011, pp. 1521–1526 (Sydney), [Online]. Available: www.ibpsa.org/proceedings/BS2011/P_1510.pdf.

[53] S. van Gemert, MPG-ENVIE: A BIM-based LCA application for embodied impact assessment during the early design stages, Master's Thesis, Eindhoven, February 2019.

[54] F. Balo. Evaluation of ecological insulation material manufacturing with analytical hierarchy process (Ekolojik yalıtım malzemesi üretiminin analitik hiyerarşi prosesi ile değerlendirilmesi), *J. Polytechnic (Politeknik Dergisi)* 20(3) (2017) 733–742.

[55] C. Eastman, P. Teicholz, R. Sacks, and K. Liston. *BIM Handbook: A Guide to Building Information Modelling for Owners, Managers, Designers, Engineers, and Contractors.* Hoboken: John Wiley, 2011.

[56] G. Hao, C. Koch, and Y. Wu. Building information modelling based building energy modelling: A review. *Appl. Energy* 238 (2019) 320–343.

[57] Tarihi Harput. www.elazig.gov.tr/tarihi-harput (Accessed February 20, 2020).

[58] Enerji Atlasi. www.enerjiatlasi.com/gunes-enerjisi-haritasi/elazig (Accessed February 12, 2020).

[59] Elazig Repa. www.yegm.gov.tr/YEKrepa/ELAZIG-REPA.pdf (Accessed February 10, 2020).

[60] Average Weather in Elazig. https://weatherspark.com/y/101227/Average-Weather-in-Elaz%C4%B1%C4%9F-Turkey-Year-Round (Accessed February 8, 2020).

[61] S. Kim and J.-H. Woo. Analysis of the differences in energy simulation results between building information modeling (BIM)-based simulation method and the detailed simulation method, in Proceedings of the Winter Simulation Conference, 2011, Winter Simulation Conference.

[62] A. Cormier, S. Robert, P. Roger, L. Stephan, and E. Wurtz. Towards a BIM-based service oriented platform: application to building energy performance simulation, Proceedings of Building Simulation 2011: 12th Conference of International Building Performance Simulation Association, 2011 (Sydney)

[63] H. Gao, C. Koch, and Y. Wu. Building information modelling based building energy modelling: A review. *Applied Energy*, 238 (2019) 320–343.

[64] L. Sanhudo, N.M.M. Ramos, J. Poças Martins, R.M.S.F. Almeida, E. Barreira, M.L. Simões, et al.. Building information modeling for energy retrofitting—A review. *Renew Sustain Energy Rev.* 89 (2018) 249–260.

[65] Z. Pezeshki, A. Soleimani, and A. Darabi. Application of BEM and using BIM database for BEM: A review. *J. Building Eng.* 23 (2019) 1–17.

[66] S. Kota, J.S. Haberl, M.J. Clayton, and W. Yan. Building Information Modeling (BIM)-based daylighting simulation and analysis. *Energy Buildings*, 81 (2014) 391–403.

[67] F. Balo. Characterization of green building materials manufactured from canola oil and natural zeolite, *J. Material Cycles Waste Manage* (JMCWM), 17 (2015) 336–349.

[68] F. Balo and H. Polat, Chapter 2. Green Design Effectiveness for a Mini Automotive-Repair Facility, in *Green Energy and Infrastructure: Securing a Sustainable Future*, Editors: J. A. Stagner and D. S-K. Ting, CRC Press/Taylor & Francis, London, 2020.

[69] H. Bahçeci and H. Polat. İnşaat Sektöründe Teknoloji Adaptasyon sorunlarının Araştırılması, *Online J. Art Design*, 8(1) 2020) 141–153.

[70] H. Bahçeci and H. Polat. İnşaat Sektöründe Yüklenici Firma Ölçeğine Göre BIM Kullanımının Araştırılması, *Online J. Art Design*, 8 (2) (2020) 124–136.

6 Limits of Waste Materials on Concrete Mixture Base Using Digital Design and Fabrication Techniques

Marwan Halabi

6.1 INTRODUCTION

Innovative developments are constantly feeding the building construction sector. By assuming this, it is possible to say that material technology is playing a vital role in the performance of contemporary buildings. Nevertheless, concrete is still used predominantly in construction. At the same time, it is an important consumer of natural resources that directly affects the environment. However, shaping and strength are some of the properties that allow designers and builders to keep on exploring and using this material to innovative extremes.

Being defined as a composite material, concrete generally comprises a matrix of aggregate, which are typically materials from rocky sources, and a binder, such as for example Portland cement, which causes the mix to hold together. There is a great diversification of concrete types, which are mainly determined by the binders' formulation and the type of aggregate used for suitability of the material's application. The characteristics of the final mix such as density, strength, and thermal resistance can be

DOI: 10.1201/9781003240129-6

affected by these variables. Aggregate on the other hand comprises normally coarse gravel or crushed rocks along with finer materials like sand.

Extensive studies have shown that a part of a common concrete mix could be replaced by some waste materials. Even though such strategy affects its performance, it still may be beneficial. Despite the potential for reduction in compression for example, thermal and acoustic properties of the concrete may improve. It's been previously and extensively tested that a limited proportion of aggregates could be replaced with, for instance, recycled plastic pigments. This would definitely increase the collection of waste, encourage its use in alternative ways, and provide possibilities for reduction of virgin natural resources use.

Therefore, a balance between performance and environmental action will be explored throughout the experiments carried out. Findings on trials will tend to indicate that the use of some waste materials in concrete can significantly contribute to a more sustainable construction industry. Further studies on environmental features such as long-term behavior of such materials in concrete and the sustainable consequences of recycling of concrete containing waste could be recommended.

6.2   WASTE MATERIALS INTO QUESTION

The three basic components of a common concrete mix are categorized as water, aggregates, and cement. One of the most popular kinds of binder is the Portland cement. This binder, which will be mixed firstly with aggregates, will have later water added in order to produce a semi-liquid paste that can be cast in order to define a form. Hydration is the chemical process that will allow the concrete to solidify and to harden. In this process, the cement and water react, thus bonding the other components together and forming a strong material. In general, the type of structure to be build conditions the mix design, in addition to the strategy for mixing, delivering it, and shaping it according to the molds or formworks.

Therefore, in such a composition such as the concrete, and from the three base components, two of them, water and binder, would be fixed, while the aggregate would provide possibilities of exploration for potential replacement with a selected series of waste materials. And it is vital to take into consideration that in addition to plastics, building construction in itself produces massive amounts of waste.

6.2.1   PLASTIC AGGREGATE

At the same time that concrete is popularly used, plastic has brought considerable benefits throughout diversified needs and use. Numerous plastic products are being consumed by the people as plastic has increasingly found many uses. However, and due to its very low biodegradability, great quantities of plastic waste provide much environmental burden. Therefore, and for the sake of both environmental protection and the economy, it is essential to develop an approach for waste management that is rational (Yang et al., 2015).

Plastics can be defined as an extensive kind of synthetic or semisynthetic materials used in a wide range of products due to their properties such as low density, high ratio of strength to weight, low cost of design and production, and unfortunately, for the planet, high durability. As of today, polymer products are still broadly used by a wide range of industries, including building and construction (Gu and Ozbakkaloglu, 2016). Regrettably, the global plastic production is not declining as it should be, if it is really decreasing at all, although some initiatives to reduce production are emerging, especially due to the fact that most types of plastics are considered to be not biodegradable in addition to chemically unreactive in the natural environment.

There are basically three means to treat plastic waste from post-consumption. According to the waste hierarchy principle, there are the landfilling, incineration, and recycling processes. Being considered the last option among the three, landfilling involves a significant amount of space and generates complications due to pollution over a long period of time. Due to the complete elimination of waste, incineration is applied in some countries. However, such process results in negative outcomes such as the release of poisonous chemicals, carbon dioxide, and toxic fly ash. Therefore, recycling would be the most appropriate solution to the tackle potential environmental concerns.

When dealing with disposing of plastic waste, there are numerous approaches for recycling management. Among them, the reuse of waste and recycled plastics in the building sector is considered to be one of the ideal ones. This method mainly involves the reuse of recycled plastics with no quality decrease throughout the service cycle. Prominently, there is potential to replace new construction materials by recycled plastics.

Extensive research dealt with the possibility of modifying conventional concrete by using recycled plastic materials. Some of the forms in which plastics have been used in concrete were in the form of plastic aggregates (PA) in the replacement of natural aggregates. Another example is the use of plastic fibers (PF), mainly for fiber-reinforced concrete applications (Saikia and de Brito, 2012).

Previous studies showed that the bulk density of PA is considerably inferior to that of typical natural aggregates and therefore it is appropriate for lightweight concrete production. Depending on the recycling method, PA may attain lower bulk density, water absorption, and melting point, but higher tensile strength. In this sense, and within a method defined as direct volume replacement, PA of similar volume could replace natural coarse aggregate or fine aggregate in concrete. Usually, concrete containing PA would be prepared, cast, and cured in similar ways to the conventional one.

In general, concrete containing PA shows lower slump when compared to conventional concrete. In relation to the concrete workability, a reduced negative effect is attained with spherical shape and smooth surface of PA rather than nonuniform plastics. Also, adding PA to the concrete leads to a reduction in density when compared to the conventional concrete, and this fact has a direct relation to the plastic aggregate replacement level. In addition, the compressive strength, elastic modulus, splitting tensile strength, and flexural strength of concrete containing PA decrease with the increase in recycled plastic aggregate (RPA). Finally, it is to be considered that according to previous studies, both water absorption and porosity of concrete containing PA increase with an increase in replacement due to the fact that plastic and natural aggregates do not mix appropriately in the concrete matrix, resulting in a porous matrix (Gu and Ozbakkaloglu, 2016).

The findings of many studies showed that concrete with RPA would not provide improvements. One of the scarce advantages was related to the plastic waste disposal. However, previous studies attempted to produce RPA using plastic and trying to solve the fillings with red sand. It was perceived that with the use of RPA instead of conventional lightweight aggregate, a reduction in compressive strength occurred. However, a range between 12 and 15 MPa of strength was attained, and this would provide enough performance for nonstructural elements (Ghataora et al., 2015). Poor bonding between plastic and cement paste is the common reason found by many studies in the decrease of the compressive strength with the increase of plastic aggregate content.

Generally, tensile failure results in failure modes in concrete. By controlling the tensile strength, it is possible to limit compressive strength losses. In this sense, by using smaller plastic particles, it is possible to reduce the compressive strength loss rather than large particles. In addition, plastic particles can be treated to provide improvement to the physical and chemical bonding with the concrete mix and reduce compressive strength losses.

6.2.2 EXPANDED POLYSTYRENE FOAM

Expanded polystyrene (EPS) is a material frequently used in the packaging sector. Some of its characteristics include high impact strength, light weight, acoustic isolation performance, and easy processing. There is a great diversity of applications with this material in fields such as for example the electrical appliances, food industry, and hardware. In the building sector, this material is used widely for heat insulation purposes.

EPS-made products can be considered to have a short service life, contrary to the ones produced from polystyrene. So in a short time, the used material is converted into residues or waste of high volume. With the purpose of reducing the problem created by the increase of such type of waste, great efforts have been made in order to avoid processes that has the potential of causing a negative environmental impact such as landfilling or incineration. In this way, this is another type of plastic recycling considerably developed lately.

Recycled EPS is a type of material that can be used, among several possibilities, not just for thermal insulation purposes in buildings, but also improve the performance of cement, for example, by reducing its porosity and permeability (Amianti and Botaro, 2008). However, its high volume to low density ratio due to the processing technology, which ranges from 10 to 25 kg/m³, is one of the main problems in EPS, increasing, for example, the transportation cost of the waste to recovery plant. Therefore, in the previous phase to recycling, it is important to reduce the EPS waste volume in order to reduce transport and storage costs in addition to decreasing the fire hazard.

There are three types of systems for disposing wastes from polystyrene foam defined as landfill, incineration, and recycling. Within the recycling, waste can go through either physical or chemical recycling processes. A high volume of EPS requires space problems in landfill, while incineration produces poisonous gas (Aciu et al., 2015). Physical recycling is characterized by the direct reuse of polystyrene wastes without the need for chemical treatment. For instance, depending on the type of polystyrene, physical recycling can be applied by grinding or crushing the foam waste, resulting only in a change of the physical form.

Studies have shown that the grinded EPS had great potential to be used within lightweight concrete mixes. One of the advantages that has been shown when using grinded recycled EPS is that it has open porosity, giving better performance possibility than using virgin EPS blobs (Kekanović et al., 2016).

The idea of developing lightweight concrete by potentially using expanded polystyrene beads is not new, and several applications are constantly being developed. It has been shown that lightweight aggregate concrete produced with EPS definitely provides benefits. However, mixing problems and weak consistency are some of the complications faced. Alternative solutions were developed but with more expensive and practice-dependent outcomes. But despite the challenges facing EPS and concrete, distinctive emphasis should be set to this product and its potential from a sustainable development perspective.

6.2.3 STONE-CUTTING SLURRY WASTE

Through the process of cutting stones into tiles or slabs of different sizes, shapes, or thicknesses, the extracted raw block is usually cut by using diamond blades. While cutting, water is spilled toward the blades to cool them down and also to absorb the dust made during the cutting process (Ammary, 2007). Besides the outcome of solid products, powder and slurry wastes in abundant quantities are produced. Besides, the waste water contains a very high alkaline level and cannot be recycled (Almeida et al, 2007). Stored in pits, the water cools until the suspended particles settle. The slurry, which becomes an inactive waste, is later collected and disposed to dry. After drying, the waste becomes an environmental hazard and starts to pollute the air with dust. Given the great amounts of waste silt continuously produced, dealing with this waste can be a significant research subject.

The water within the stone-cutting slurry waste can lead to considerable environmental problems. For instance, dumped areas with slurry, which contains stone powder in abundancy, may not support vegetation, besides triggering air pollution. Depending on the stone type and process of waste generation, different physical and chemical properties can be generated in the stone-cutting slurry waste (Arslan et al., 2005).

Discharged in open areas and causing the possibility of serious problems related to environmental pollution are the main drawbacks of the usual management of marble and stone-cutting industries, which usually dispose their waste in open areas. However, if such type of waste could be recycled in any way possible, then problems of stone waste disposal could be drastically reduced (Demirel, 2010).

In previous trials, marble powder and fly ash were applied as partial replacement of cement in self-compacting concrete mix, a type of mixture that flows beneath its own weight and without the need for external vibration to undergo compaction (Pala et al., 2015). The filling and passing ability results improved by the use of 10 percent waste marble powder and 25 percent fly ash. The possibility of designing self-compacting concrete mix with different substances of local stone-cut powder partially replacing fine silica aggregate was detected. With increase of slurry waste content, it has also been observed that the slump flow decreased (Suliman et al., 2017).

Besides drying concerns, the chemical composition variations of the slurry have the potential to offer complications in respect to quality control. Another drawback is the volume of slurry from several sources, which can be excessive for an entire reuse. This may result in problems concerning waste management. In addition, the chemical composition variations may affect both production and the quality of the final products (Chang et al., 2010).

6.3   CASE STUDY

Previous studies showed interesting examples of dealing with waste materials and concrete. Previous trials have shown that it is somehow manageable and possible to integrate, for example, plastics or stone-cutting slurry waste into the concrete mixture. However, the idea is to insert not just one type but a series of waste materials and in the greatest proportions possible to produce a concrete that would have enough properties to serve as a material for the production of urban furniture. Initial assumptions were based on the use of recycled plastic particles, grinded polystyrene foam residues, stone-cutting slurry waste, and sawdust from woodwork, although preliminary trials assessed for the removal of components from wood due to their negative effect on the concrete curing process.

The experiment of developing the material was linked to design and fabrication divided into the following modules: design and simulation, prototyping, and testing mixtures with waste materials.

6.3.1   DESIGN AND SIMULATION: OUTDOOR BENCH

In parallel with the material test, it was necessary to design an object in order to test the material performance at different levels, especially when dealing with unconventional shapes. The design in question would have certain characteristics that could put into test the compressive strength of the material, its durability related to diverse weather conditions, and aesthetics, with the possibility of introducing, for example, texture or colors. The form decided to be designed was an outdoor chair with single curvature and extrusion. The shape of the chair would provide certain visual dynamism, but at the same time some degree of instability due to its exaggerated curvature. However, such bend would put into test the performance of the material related to nonstandard shapes, such as, for example, its compressive strength.

Topology, which is mainly concerned with the properties of a geometric object that are preserved under continuous deformations, is a common strategy used when designing with the help of digital technologies. However, finding a balance between the limits of what a material can perform in relation to its form could imply in a compromise for the sake of the environment. Complex shapes of any computational design principle probably could not be taken to extremes. Nevertheless, interesting shapes could come out of the process. However, during the first stage, a single curved surface was designed in order to provide a bench of bended legs to examine the

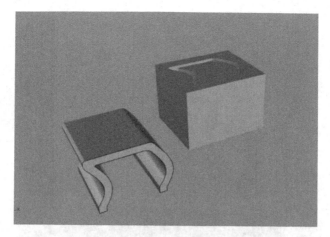

FIGURE 6.1 Design of a curved bench of 40 cm × 50 cm × 40 cm to be fabricated as a mold for casting.

performance of one of the mixtures developed in parallel (Figure 6.1). The prototype would have 50 cm of width by 40 cm of depth and would be 40 cm high.

6.3.2 PROTOTYPING: MOLD FABRICATION

Once the design was defined, a strategy for the fabrication of casting mold had to be devised. The mold for casting would be produced by using 10 cm foam board leftovers from previous works and stacked together to provide a bench depth of 50 cm, in addition to 12 mm thick plywood leftovers for the provision of structural stability of the mold, which would be later assembled by using 6 mm steel rods and nuts (Figure 6.2).

The 5-cm-thick low-density polystyrene boards would be cut by using a CNC milling machine with a tool bit of 12 mm. The cut foam panels would be later superimposed over each other and without any adhesives, only the steel rods so that it would be easier to dismantle the mold after casting, besides that the foam is quite fragile material for the casting process.

6.3.3 TESTING MIXTURES WITH WASTE MATERIALS

When dealing with several waste materials, the first step in the process was to find the ideal composition that, by modifying a standard concrete mixture, would allow for a combination that could stand certain amount of stress. Therefore, cylinders of 10 cm diameter and 20 cm height were to be cast in order to be later taken to compression test. The first mixture started with a conventional concrete mix of one part cement, two parts sand, and three parts aggregate ratio.

During the first trials, the replacement of aggregates would be arbitrarily applied, and once having observed their preliminary properties such as stiffness and compactness, the mixtures would be applied and recorded. For example, instead of having one part cement, two parts sand, and three parts aggregate, the mixture number two

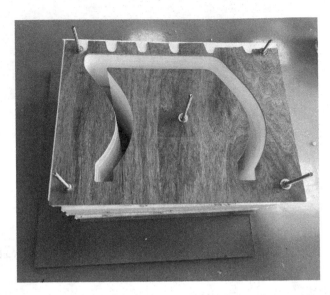

FIGURE 6.2   Fabrication of the mold for casting using leftovers from previous tasks.

would have one part cement, one part sand, two parts gravel, and one part grinded polystyrene foam. At a later stage, proportional replacement was carried out in order to produce cylinders for test. A total of 18 mixtures were prepared (Table 6.1).

During the process, recycled plastic pigments from high-density polyethylene (HDPE) were used. It is considered to be one of the most common plastics used by several industries. Some of its characteristics include being lightweight, relatively transparent, water and temperature resistant, in addition to having high tensile strength.

Besides recycled plastic, grinded polystyrene foam, mainly from EPS, was used in addition to the stone-cutting slurry waste mainly from marble. Grinded polystyrene foam particles between 4 and 8 mm were used in the dry mixing composed of gravel, sand, cement, polystyrene, and sawdust, with water added later.

After the casting and curing of all the cylinders, and due to failure of some samples, which showed fragility in line with either exaggerated amounts of waste or lack of proper bonding such as the case of the sawdust added to the mix, a selection has been made in order to test their compression (Figure 6.3).

One of the drawbacks of the mixtures was their compressive strength, which was far from the conventional concrete mix (Table 6.2). However, since the intention was to use the mix for purposes requiring relatively low compression, some of the mixtures could be tested in the casting test of the bench. On the other hand, one of the curious observations during the test was the result produced by one of the mixtures, defined as mix number 4, which had aggregate considerably reduced and sand completely replaced by recycled plastic and polystyrene pigments, where the initial weight of the cylinder was almost six times lighter than the conventional mix. After 28 days, its final weight was 44.8 percent compared to the conventional mix, which is around 3.42 kg (Table 6.3).

TABLE 6.1
Concrete Proportions Using a Series of Combinations for Test, Where One of the Most Interesting Mixtures Was Number 4 Due to Its Texture and Very Light Weight

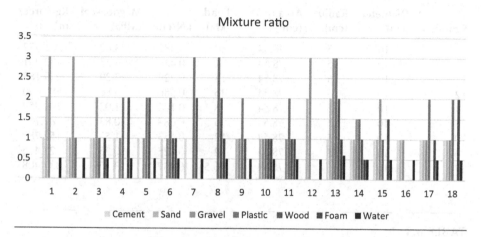

FIGURE 6.3 Sample of cylinders for the stress test after the curing process, comprising of different physical characteristics and weights.

Later, the mixture number 4 would be developed as mixture number 18, with half part of stone-cutting slurry waste added, and later mixture 21, with one part of the slurry. This component would act as a filler, providing more smoothness for the form in addition to a much greater compressive resistance, even though its weight considerably increased. Another advantage of this mixture was its percentage of modified

TABLE 6.2
Compressive Test Results that Shows Low Compressive Performances in Comparison to the Standard Concrete

Sample #	Diameter (cm)	Radius (cm)	Area = Π*r² (cm²)	Load (kN)	kN/cm²	Megapascal (MPa)	Kg-Force/ m²
9	10	5	78.54	30	0.382	3.82	381,972
3	10	5	78.54	34	0.433	4.33	432,901
18	10	5	78.54	18	0.229	2.29	229,183
12	10	5	78.54	46	0.586	5.86	585,690
4	10	5	78.54	8	0.102	1.02	101,859
13	10	5	78.54	6.5	0.083	0.83	82,761
10	10	5	78.54	12	0.153	1.53	152,789
6	10	5	78.54	2	0.025	0.25	25,465
17	10	5	78.54	26.5	0.337	3.37	337,408

TABLE 6.3
Combination of Material Mixes and Their Posterior Weight Results After Curing, Noting that on the Cement Base, "w" Stands for White Cement

Mix number	Cement	Sand	Gravel	Plastic	Wood	Foam	Water	Volume (m3)	Weight (gr)	Stone waste powder
1	1	2	3	0	0	0	0.5	0.00157	3,746	0
2	1	1	3	1	0	0	0.5	0.00157	1,126	0
3	1	1	2	1	0	1	0.5	0.00157	2,638	0
4	1	0	1	2	0	2	0.5	0.00157	1,502	0
5	1	0	1	2	2	0	0.5	0.00157	949	0
6	1	0	1	2	1	1	0.5	0.00157	1,546	0
7	1	0	0	3	2	0	0.5	0.00157	725	0
8	1w	0	0	3	2	1	0.5	0.00157	fail	0
9	1w	1	1	2	1	0	0.5	0.00157	2,256	0
10	1w	1	1	1	1	1	0.5	0.00157	2,105	0
11	1w	0	1	2	1	1	0.5	0.00157	fail	0
12	1w	2	3	0	0	0	0.5	0.00157	3,380	0
13	1	2	3	3	2	1	0.6	0.00157	1,926	0
14	1	1	1.5	1.5	1	0.5	0.5	0.00157	fail	0
15	1	1	2	1	0	1.5	0.5	0.00157	1,862	0
16	1	1	1	0	0	0	0.5	0.00157	1,400	0.25
17	1	1	1	2	0	1	0.5	0.00157	2,264	1
18	1	1	1	2	0	2	0.5	0.00157	1,975	0.5

composition, where 50 percent of it was from waste material sources and the other half from virgin ones.

Mixture 4 results led to some questions on whether such mixture would resist the typical outdoor conditions and the can be used for a piece of urban furniture. Therefore, a production of the prototype of the bench previously mentioned was carried out using both mixtures number 4 and 21 and were cast using the same mold. In both cases, the benches would be reinforced by using 6 mm steel bars due to the considerable load that the bench was supposed to carry (Figure 6.4). According to the previous tests, the bench with the mixture number 4 would have a porous texture while the number 21 would have a completely smooth texture very close to the ordinary concrete and due to the use of stone-cutting slurry waste. The relative difference in weight was also expected to be noted, even though both mixtures would end up being lighter than ordinary concrete mix.

It has been generally perceived that the compressive and tensile strengths decrease when replacing aggregates by plastic in a concrete mix. This is mainly due to the weak bond among the mix components, especially the plastic. In addition, constant use and external conditions may cause certain types of damage. However, both prototypes have been exposed in public areas for public use, in addition to being exposed to different weather conditions during a 12-month period, being noted that they still remain in their original physical states (Figure 6.5).

6.4 CONTRIBUTION TO THE SUSTAINABLE DEVELOPMENT GOAL 9

The environment and the quality of life are being directly affected by the global crisis caused by environmental damage. In order to face this challenge, the United Nations adopted the Sustainable Development Goals (SDGs), an agenda that contains 17 objectives of universal application (Guterres, 2019). The definition of sustainable development stands as without conceding the ability of future generations to meet their own necessities, to develop capabilities of meeting the needs of the present.

Since the start of 2016, countries were challenged to put in efforts in order to achieve a sustainable world by 2030 (UN, 2015). Such development requires concentrated efforts to build an inclusive, sustainable, and resilient future for people and the planet (Bassi et al., 2019). In order to achieve sustainable development, a vigorous coordination among economic growth, social inclusion, and environmental protection is essential, since these bases are vital and interconnected to the people's well-being.

One of the SDGs that has direct link to the case study explained is SDG 9. This objective has to cover industry, innovation, and infrastructure. They are the base for a series of targets within the goal. However, this experiment needs to emphasize on the intention to act in a way that a sustainable development becomes a means of thinking and working to solve problems that contribute to environmental issues, without thinking of its implication scale. In this sense, initiatives are important for awareness and engagement in actions.

Aside from commitment of governmental engagement, there must be a deep involvement from the educational field. Academic institutions should be boosted, and somehow it was achieved after the case study provided the possibility of hosting an

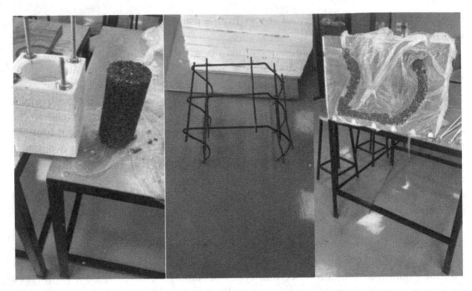

FIGURE 6.4 Mixture number 4 has a porous effect, is light weight, and was reinforced using 6 mm steel bars before casting. The mold was covered with a thin sheet of nylon for its protection and to attain a smooth surface.

FIGURE 6.5 Bench from mixture number 4 (left) and number 5 (right) after a 12-month period testing under different outdoor weather conditions.

international node of collaboration developed for students and professionals as an initiative to be explored. Since the SDGs comprises challenges at the universal scale, this can inspire collaboration at the cross-cultural levels, aiming at attaining positive outcomes in a more effective manner. If the suitable recommendations for the SDGs are to be delivered by such research, a cooperative development among academic faculties and professional organizations is certainly required. Despite the activists' vital role in generating awareness, a level of maturity must be established in order to take proper and innovative actions for the coming generations.

6.5 CONCLUSIONS

One of the intentions of the experiment was to take the advantage of waste materials without the need to go through any type of chemical processes, assuming that it was either waste for disposal or material for recycling, executing replacements within the conventional concrete mix, and ending up with a matrix that had enough characteristics to be used in the production of possible urban furniture and outdoor components.

The advantage of the study is that the formal approach can be supported by advanced design and fabrication techniques, where it would be possible to produce a series of customized components to be adapted according to the context. However, unconventional design would be balanced with the aim of the study, which was not to attain a structural concrete of high compression strength, but to attain enough strength that could help in the production of a component that could bear as much waste material as possible.

Within the case study, the drastic modification of mixtures affected water absorption, in addition to the concrete porosity having recycled plastic. This was observed during the mixing process and some mixes, especially the ones without cutting-stone slurry waste, resulted in porous mixes. In addition, the increase of recycled EPS directly resulted in the decrease of the mechanical characteristics of the concrete mixes even though its workability increased and density decreased. Experimental data highlighted that recycled EPS influenced the characteristics of the concrete mixes.

The spreading of recycled EPS in concrete mixtures was not even, and it was visible that the granules had the tendency to flow, especially when applying high dosages such as in mix number 4. Nevertheless, the experiment was able to sustain concrete with polystyrene granules as a concrete mix with relatively reduced weight from the conventional one, and with the potential of being used as a material with nonstructural purposes. The case study showed that there is a considerable potential to develop mixtures with not only recycled plastics, but also with other waste materials such as the stone-cut slurry waste. However, some of the parameters to be taken into consideration during further studies should be related to stress, strain, rigidity, and deformation, in addition to temperature effects.

It was usually appreciated during the preliminary tests that replacing plastic into a concrete mixture caused a negative effect in the compressive and tensile properties resulting from the weak bond between the plastic and mixture components. At the same time, it was confirmed that the plastic aggregate of sufficiently small sizes to avoid failure was very effective. The stone-cutting slurry waste would then play a significant part in the binding process, filling the gaps of the porous mix.

The advantages of employing digital tools during the case study were found not only on hardware and software principles, but also on blending of ideologies with the aim of trying to develop sustainable materials with the support of advanced procedures. The case study showed that the experiment has the potential to motivate students, fresh graduates, and professionals to come up with innovative ideas for further progressive development. The intentions of the study had the potential to cover some of the challenges related to the SDG9 and also promoting possible stakeholder commitment. Nevertheless, preparation over advanced disciplinary education should be boosted for the promotion of an inclusive and sustainable industrialization.

Some of the intended learning outcomes of the experiment are related to the ability of investigation and appraisal in the selection of an alternative constructional and material systems relevant to architectural design. The experiments dealt with the recognition of basic physical properties of building materials, components, systems, and their environmental impact. Adapting ideas about waste with the help of innovation, art, and technology is a way to uncover possibilities in working with waste and recycled materials, particularly when evolving approaches that intend to provide a sense of consciousness through exploration. This is only done in expectation to encourage people to act upon sustainable principles.

REFERENCES

Aciu, C., Manea, D.L., Molnar, L., and Jumate, E. (2015). Recycling of Polystyrene Waste in the Composition of Ecological Mortars. *Procedia Technology*, 19: 498–505.

Almeida, N., Branco, F., and Santos, J. R. (2007). Recycling of Stone Slurry in Industrial Activities: Application to Concrete Mixtures. *Building and Environment*, 42: 810–819. doi:10.1016/j.buildenv.2005.09.018

Amianti, M., and Botaro, V. R. (2008). Recycling of EPS: A New Methodology for Production of Concrete Impregnated with Polystyrene (CIP). *Cement Concrete Comp*, 30: 23–28. doi: 10.1016/j.cemconcomp.2007.05.014

Ammary, B. (2007). Clean Cutting Stone Industry. *International Journal of Environment and Waste Management*, 1: 106–112. doi:10.1504/IJEWM.2007.013627

Arslan E. I., Aslan, S., Ipek, U., Altun, S., and Yaziciolu, S. (2005). Physico-Chemical Treatment of Marble Processing Wastewater and the Recycling of Its Sludge. *Waste Management & Research*, 23: 550–559. doi: 10.1177/0734242X05059668.

Bassi, A., Casier, L., Laborde, D., Linsen, M., Manley, D., Maennling, N., Mann, H., Siersted, M., Smaller, C., Steel, I., Uzsoki, D., and West, J. (2019). *Modelling for a Sustainable Development: New Decisions for a New Age*. Manitoba: The International Institute for Sustainable Development (IISD).

Chang, F. C., Lee, M. Y., Lo, S. L. L., and Lin, J. D. (2010). Artificial Aggregate Made from Waste Stone Sludge and Waste Silt. *Journal of Environmental Management*, 91: 2289–2294. doi:10.1016/j.jenvman.2010.06.011

Demirel, B. (2010). The Effect of the Using Waste Marble Dust as Fines and on the Mechanical Properties of the Concrete. *International Journal of the Physical Sciences*, 5(9): 1372–1380.

Ghataora, G., Alqahtani, F., Khan, M., Dirar, S., and Al-Otaibi, A. (2015). Lightweight Concrete Containing Recycled Plastic Aggregates. International Conference on Electromechanical Control Technology and Transportation (ICECTT 2015).

Gu, L., and Ozbakkaloglu, T. (2016). Use of Recycled Plastics in Concrete: A Critical Review. *Waste Management*, 51: 19–42. doi: 10.1016/j.wasman.2016.03.005.

Guterres, A. (2019). *Report of the Secretary-General on SDG Progress 2019—Special Edition*. United Nations. Retrieved on September 10, 2019, from: www.un.org/sustainabledevelopment/progress-report/

Kekanović, M., Kukaras, D., Ceh, A., and Karaman, G. (2016). Lightweight Concrete with Recycled Ground Expanded Polystyrene Aggregate. *Technical Gazette*, 21, 2: 309–310.

Pala, K.P., Dhandha, K. J., and Nimodiya, P. N. (2015). Use of Marble Powder and Fly Ash in Self Compacting Concrete. *International Journal for Innovative Research in Science & Technology*, 1: 475–479.

Saikia, N., and de Brito, J. (2012). Use of Plastic Waste as Aggregate in Cement Mortar and Concrete Preparation: A Review. *Construction Building Matter,* 34: 385–401. doi: 10.1016/j.conbuildmat.2012.02.066.

Saikia, N., and de Brito, J. (2014). Mechanical Properties and Abrasion Behaviour of Concrete Containing Shredded PET Bottle Waste as a Partial Substitution of Natural Aggregate. *Construction Building Matter*, 52: 236–244. doi: 10.1016/j.conbuildmat.2013.11.049.

Suliman, M., Alsharie, H., Yahia, Y., and Masoud, T. (2017). Effects of Stone Cutting Powder (Al-Khamkha) on the Properties of Self-Compacting Concrete. *World Journal of Engineering and Technology*, 5(4): 613–625. doi:10.4236/wjet.2017.54052.

United Nations (UN) (2015). *Transforming Our World: The 2030 Agenda for Sustainable Development*, 4–10. Retrieved on July 25, 2019, from: https://sustainable development.un.org/content/documents/21252030%20Agenda%20for%20 Sustainable%20Development%20web.pdf

Yang, S., Yue, X., Liu, X., and Tong, Y. (2015). Properties of Self-Compacting Lightweight Concrete Containing Recycled Plastic Particles. *Construction and Building Materials*, 84: 444–453. doi: 10.1016/j.conbuildmat.2015.03.038.

7 Remote 3D Printing for the Integration of Clay-based Materials in Sustainable Architectural Fabrication

Yomna K. Abdallah, Secil Afsar,
Alberto T. Estévez, and Oleg Popov

7.1   INTRODUCTION

Managing the remote 3D printing process requires a concrete understanding of cyber manufacturing concepts and terms, including big data, smart analytics, Internet of Things, cloud-based/distributed computing, cyber security, and all the parameters that are included in each of these. "Big Data" is the main heterogeneous mass of digital data produced by design and manufacturing entities whose characteristics

DOI: 10.1201/9781003240129-7

(large volume, different forms, speed of processing) require specific and increasingly sophisticated computer storage and analysis tools (Riahi and Riahi, 2018), while the Internet of Things is type of network to connect anything with the internet based on stipulated protocols through information sensing equipments to conduct information exchange and communications in order to achieve smart recognitions, positioning, tracing, monitoring, and administration (Patel and Patel, 2016). These two are the main parameters in the cloud computing or distributed computing where the cloud plays an important role in organizating and maintaining huge data within limited resources. Cloud facilitates resource sharing through some specific virtual machines provided by the cloud service provider (Divakarla and Kumari, 2010). This distributed computing requires security of the data, and the assets that produce or assess it. Thus cyber security is needed, and cyber security manages the set of techniques used to save the integrity of networks, programs and data from unauthorized access (Seemma et al., 2018). All these aspects contribute to smart manufacturing (SM), which is moving manufacturing practices toward the integration of physical and cyber capabilities and taking advantage of advanced information for increased flexibility and adaptability. SM is often equated with "Industry 4.0" (Lee et al., 2014). Managing the 3D reomte printing proess also depends on the understanding of the interdependencies between the parameters that rule this process, as shown in Figure 7.1, a diagram of the design to production cyber manufacturing optimization.

As inferred from the diagram, the remote 3D printing process is to be considered the most sophisticated manufacturing process in the cyber manufacturing field, especially when it involves nonstandard material synthesis and calibration. However, before diving into the detailed description and analysis of this process through this case study, the first an essential overview of the cyber manufacturing terms, elements, and processes are presented to provide a concrete basis.

7.2 CYBER MANUFACTURING

Cyber manufacturing includes the translation of data from interconnected manufacturing systems into predictive and precise operations to achieve optimized remote production processes. It combines industrial big data and smart analytics to discover and comprehend visible and invisible production issues for decision making (Lee et al., 2016). Cyber manufacturing evolved from e-manufacturing, which enables manufacturing operations to be successfully conducted remotely through the use of tether-free (i.e., Internet, wireless, Web, and so on) communication and predictive technologies (Lee et al., 2016; Lee, 2003). Both e-manufacturing and cyber manufacturing aim to reduce unexpected production errors and time. However, cyber manufacturing is a more complex and data-rich system, which integrates smart analytics, distributed systems, control science, and operation management to construct a cyberphysical model (Lee et al. 2015).

Cyber-physical systems (CPS) manage interconnected systems that are composed of physical assets and cyber networking to enable the gathered manufacturing information from all related perspectives to be closely monitored and synchronized between the physical workshop assets and the cyber computational space. Thus, the

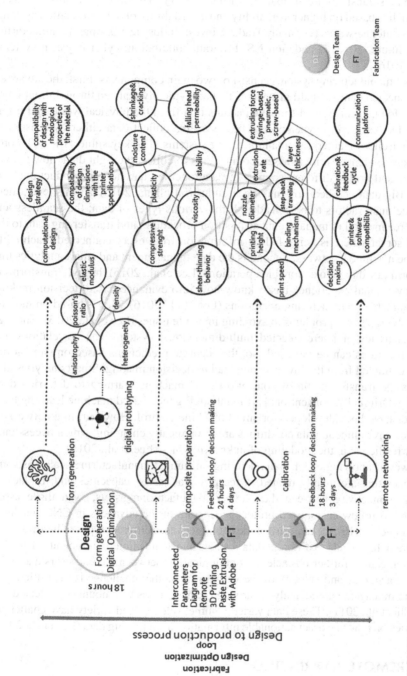

FIGURE 7.1   The parameters and interdependencies of the remote 3D printing process.

production process can be performed more efficiently, collaboratively, and resiliently. This trend is transforming the manufacturing industry to the next generation, Industry 4.0, which is based on higher availability and affordability of data acquisition systems and computer networks (Germany Trade & Invest. Industrie 4.0: smart manufacturing for the future, 2016; Foundation NS. Program solicitation: cyber-physical systems (CPS), 2016).

Cyber manufacturing systems consist of two main components. First, the advanced interconnected data acquisition systems that are typically realized through Internet of Things (IoT), ensuring real-time data acquisition from the physical world and information feedback from cyberspace. These systems collect data either by sensors or by a controller or enterprise manufacturing systems. The key subjects in IoT comprise how to identify critical assets/components to collect the right data, how to synchronize and bridge different sources of data together, and how to conduct analysis (ABB. Big data and decision-making in industrial plants, 2016). At this level, there are three main factors to be considered: the various types of data; the seamless and tether-free method to manage data acquisition procedure; and transferring data to the central server where is being fed by information from every connected machine to form the machine's network. Second is the data management and smart analytics that transform raw data into actionable operations (Lee et al., 2015) through transforming the conventional experience-based know-how into evidence-based decision making for sustainable manufacturing operations (Lee et al., 2016). This is based on moving from solving visible problems to avoiding invisible issues (Lee et al., 2013). Through smart analytics of interconnected multidimensional systems, the correlations and causal functions can be modeled. So, the accurate predictions based on simulations can be extracted from the raw data and lead to decision making. Smart analytics also enables the transformation of control-oriented machine learning to data-rich deep learning. Originally, conventional artificial intelligence-based machine learning technologies have been developed for smart machine control. However, in a networked, data-rich environment, data conditions are dynamically changing, which necessitates greater resilience in the modeling of unknown issues (Lee et al., 2013).

However, applying cyber manufacturing on limited manufacturing scale has some challenges that are mainly related to the physical assets' capacities and the lack of unified connectivity between these cyber manufacturing assets, even those assets that tend to follow customized protocols. In spite of the advances in CNC machines (MTConnect 2016), the majority of the equipment uses different types of hardware and software, which leads to different data formats and acquisition requirements. This data mismatch causes further obstacle in communication between the end-users and holds back the accuracy and velocity of the workflow. Another challenge is the difficulty in big data management and analysis, arising from the massive amount of collected raw data (Shi et al., 2011). These data varied volume, velocity, and variety have challenged industries on how to extract actionable information from this big data (Lee et al., 2013).

7.3   REMOTE 3D PRINTING

Remote 3D printing is one of the manufacturing processes that are included in cyber-physical manufacturing. However, this process is more complicated than the simple subtractive fabrication techniques, such as CNC milling. The complication of the remote 3D printing process is emerging from direct and indirect relation of all the printing process parameters, such as the printer specifications, the printing material composition, design, and the printing environment including printing temperature and humidity levels. The process gets even more complicated in case of experimenting with nonstandard materials that require accurate composition and multiple calibration-feedback loops that affect the optimization of the material's mechanical properties through the fine-tuning of the mixture, design, printing time, and the potential of replicating this print in the mass-production scale. It is important to highlight the effect of the printing environment, as the ambient temperature and humidity levels control the evaporation and the desiccation rate of the calibrated materials and the overall solidification quality. As 3D printing is scale-dependent, the printing speed affects the printing time, which controls the efficiency of mass production of the printed design. All these attributes add difficulty in managing the 3D printing process remotely and require prompt and efficient responses between the remotely distributed collaborators. Thus, the difficulty of managing this remote manufacturing process increases relatively with the increase in the number of collaborators and their geographical distribution. Post-manufacturing and distribution are other debatable issues. The difference in climatic conditions of the location of manufacturing to the user's location has a significant effect on product performance and durability in the operative environment. Thus, all these interconnected aspects should be considered in the design and simulation phase as well as the continuous interplay of calibration-feedback between these interdependencies to optimize the production process.

7.4 MATERIALS AND METHODS (CASE STUDY)

In this study, the democratized and fragmented cyber-physical manufacturing process is optimized to convert this technology to a reachable level. This enables startups and limited scale manufacturers to benefit from this advancement. Also, it provides a handy solution for researchers and academic entities that might lack the financial capacity to support establishing their own fabrication labs or the capacity to collaborate with large-scale manufacturers. This point is crucial at forming all the aspects and assets of the cyber/physical manufacturing process, limiting the scale of networking and the number and capacity of the machines to the most available and easily affordable by the emerging startups. In the particular case of this study, remote 3D printing on the architectural scale is investigated. This manufacturing process involved more sophisticated phases as it included synthesizing and calibrating nonstandard clay-based materials, which exposed a number of interconnected and interdependent problems that are related to each of the parameters that are forming the basis of this process. In this section, the assets that are involved in the remote 3D printing process, in an attempt to simplify the practice for researchers and designers from the architectural background, are categorized.

7.4.1 DESIGN (3D PRINTING ON ARCHITECTURAL SCALE)

The main objective of the experimentation is attaining sustainability in architecture from design to production, and in performance. Thus, this attempt to use 3D printing on an architectural scale to print environmentally and climatically sustainable architectural facades are aimed to benefit from the numerical precision of the 3D printers. For achieving this, the design approach moves away from the solids and head toward the lines and voids, inspired by biological structures of corals that uses the minimum materials to build its self-bearing structural membranes without hindering its mechanical properties and shear resistance (see Figure 7.2 of *Mycetophyllia danaana* Caribbean coral structure that shows deep valleys convoluted from the original point of growth).

The mathematical simplification and form generation of such system, as coral structures, are based on ramifications of divisions of XY lines/4 from each of its initial points or lines of growth and expand exponentially following a fractal mathematical rule that if a branch has one line. These coral structures like other structures found in nature tend to grow by following a random walk pattern. However, they always follow specific attractors that affect their growth during their random walk. These attractors could be sources of nutrients or growth conditions such as sunlight or moving water energy due to static and dynamic forces.

Following these coral structures, the design of an architectural façade unit of 25 × 50 × 25 cm was designed following these steps. Through rebuilding and dividing a reference curve, the generated line segments are grown randomly but with specific rules to control the shape. In order to avoid collision between these growing lines and to control the distance that is adjusted based on the desired line thickness of the 3D print, the spheres from line points are developed. Figure 7.3 shows the parametric form generation in grasshopper 3D.

The grasshopper plug-in Kangaroo allows generating physical strength to grow the curves that are stimulating the inertia with respect to developed spheres. Besides, the Anemone plug-in controls the iterations and helps to produce steadier fractal patterns. In the end, the generated form is manipulated to generate the final design. In this design, even though clay as a fine-grained rock or soil is a static element of nature with inertia zero, by applying differential growth to a curve along a 3D surface, the design is attributing the growth mechanism in nature to clay. In Figure 7.4, the preparation of the design model to production is exhibited.

FIGURE 7.2 Coral membrane structure of *Mycetophyllia danaana*. Left, photograph by Nicole Helgas. Center and right, showing deep valleys meandering from the original point of growth. Photograph by Mary Stafford-Smith and Veron archives.

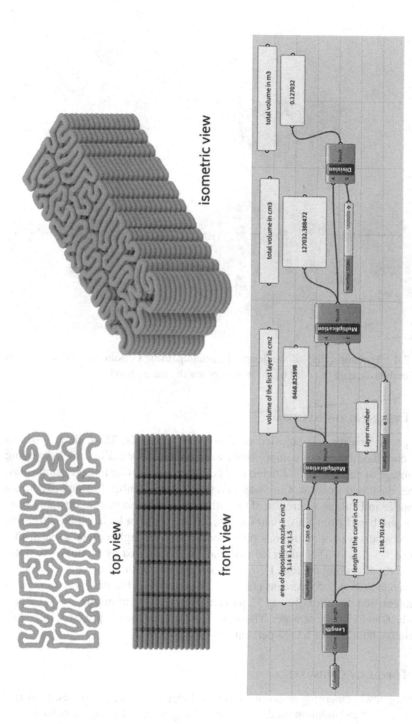

FIGURE 7.3 The design of the façade building unit, and the calculation of the estimated volume, developed by Rhinoceros 3D, Grasshopper (by the authors).

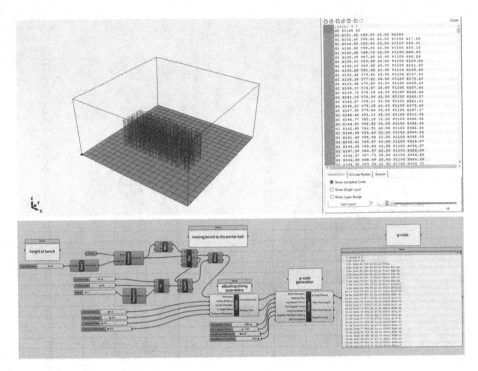

FIGURE 7.4 The preparation of the design for printing, using Grasshopper 3D to perform slicing of each layer, data provided by the manufacturer (by the authors).

7.4.2 DESIGNING THE MATERIAL

The design materiality was adjusted to clay-based materials. Thus, three types of different clay-based material compositions were prepared for calibration by varying the percentage of each of the following raw components: the Ukrainian ceramic clay, the marble dust, and sawdust. Clay was purchased from Donbas Ceramic Bodies, a local supplier in Ukraine where the 3D printing occurred, the marble dust and saw-dust were purchased from Kiev Plant Granite, which is also a local supplier in Kiev, Ukraine. Sawdust was of coniferous pine wood, sifted through a metal sieve with a diameter of 1.5–2 mm. The calibrated materials were prepared by mixing with water as follows: the first composition (1) was 75 percent clay with 25 percent water content, the second (2) was 55 percent clay and 27 percent marble dust and 18 percent water content, and the third (3) was 55 percent clay, 23 percent marble dust, 2.2 percent sawdust, and 17 percent water. These different mixtures were prepared in 5–7°C, and in relative humidity of 65–75 percent.

7.4.3 TOPOLOGY OPTIMIZATION

Considering newly emerging materials, bio-based materials, or composites, it is difficult to use existing simulation models due to missing data and complex behaviors of the materials like high water permeability, anisotropy, and heterogeneity. In general, a combination of finite element method (FEM) and fabrication investigations based on mechanical testing, microscopic imaging, and finite element modeling can be used for collecting data from specimens and predicting the mechanical properties along with its deformation behavior. However, due to restrictions during the lockdown, the mentioned tests could not be carried out for this research, and FEM models are generated with several assumptions.

From the ASTM standards and previous researches, some of the data like Poisson's ratio, Young's modulus, density, and compressive strength are determined by assuming the composite as the soft soil. Another assumption is considering sawdust and marble dust are dispersed homogeneously into the composite and generating estimated results based on isotropic material behaviors. Later, by defining fixed supports and forces, simulation is run in ANSYS Workbench 19.2 to visualize the total deformation and equivalent stress models. In the end, the results for optimization region and topology optimization are obtained, respectively.

7.4.4 3D Printer and Calibration Process

The 3D printer used in this study, with dimensions of height in 225 cm, length in 295 cm, and depth in 25 cm has a paste extruder that has an air pressurized piston.

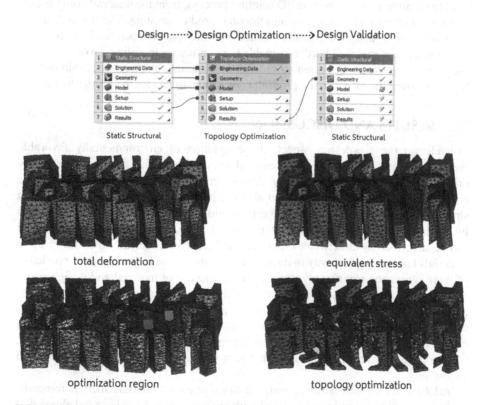

FIGURE 7.5 ANSYS Workbench 19.2 to visualize the total deformation, equivalent stress, optimization region, and topology optimization.

The nozzle diameter was 1 cm. The printing was carried out at an ambient temperature of 5–7°C, and in relative humidity of 65–75 percent. The printing speed was 10 mm/s.

The calibration process started with line testing of the different three materials compositions 1, 2, and 3, which were mentioned in the materiality section. The Simplify3d cutting and printing software was used by the manufacturer.

7.4.5   REMOTE 3D PRINTING (COMMUNICATION, FEEDBACK-DECISION MAKING)

In this study, the collaboration was between a design/research entity, which is iBAG-Institute of Biodigital Architecture and Genetics, Universitat Internacional de Catalunya (UIC), which is located in Barcelona, Spain, and the manufacturer collaborator located in Kiev, Ukraine. The communication between the collaborators was carried on using the email and texting app of WhatsApp. The designed model was shared on the cloud between the collaborators through networking by linking the iBAG designer's computer to the computer linked to the printer machine. However, the designs of the manufacturer were included later in the network in order to enable prompt problem solving to the urgent issues, which weren't attended immediately by the designer team of iBAG due to varied working hours. Snapshots and videos were used to monitor the entire remote 3D printing process, from the material composition stage, through the calibration, and into the final product printing. The iBAG designer used parametric design platforms Rhinoceros 3D+grasshopper, and its plug-ins Anemone, and Kangaroo in order to enable the continuous modification of the design model in any stage from design to production. The manufacturer used Simplify3d for transferring the model to the printing code compatible with the printer machine.

7.5   RESULTS AND DISCUSSION

Considering the eco-design aspect, the adaptation of environmentally favorable materials and its fabrication techniques need an interdisciplinary thinking approach in order to lead to further developments. Among contemporary architects, not only the emergency call of climate change but also the increased tendency toward free-form structures leads to a rise in the popularity of bio-based and biodegradable materials under the discourse of the "new materiality." In spite of this growing trend, which can be described as in between the urgent and emergent, the full potential of these materials has been tapped only in the rarest cases due to the lack of proper knowledge of their fabrication techniques, along with the absence of material and design simulation techniques. Bio-based and biodegradable materials mostly have weak mechanical properties, high water permeability, and due to anisotropy and heterogeneous features, their behaviors are difficult to predict. In this regard, although developing composites is a promising way of overcoming the weaknesses of these materials, the compositions and processing techniques should be engineered and calibrated extensively.

Adobe is the earliest building material made of earth materials such as minerals, rocks, soil, and water and can be mixed with organic matter. Clay is a vital element of adobe with its wet plastic properties.

Due to clay minerals, which also affect its particle size and geometry, its mechanical properties distinguish it from other earth materials like silt and sand. While clay helps in shape-preserving post drying, gravel, silt, or sand give the print the required strength. Clay can appear in various colors, from white to gray or orange to brown. The mechanical behaviors of sun-dried adobe and fired adobe alter significantly. The sun-dried adobe structures, which are typically used in dry climates, show high durability, abundance, economic and sustainable features, as well as providing a greater thermal mass and better indoor air quality.

Thanks to the increase in the awareness of the construction sector of its ecological footprint, possible stabilizing materials for adobe have been searching as an alternative to conventional cement and lime materials. Nowadays, fibers, biopolymers, and byproduct solid materials (wastes) are used to advance the geotechnical properties of clay. In this research, high clay content adobe is developed by using marble dust, sawdust, and water.

The physical prototyping steps to advance mechanical properties and stability of biocomposites can be considered time- and material-consuming processes. At this point using digital prototyping techniques such as generating finite element analysis simulation to optimize material, form, and digital fabrication tools can help to adopt these materials in architecture and interior design elements on a mass-production scale.

In this section, the results of this case study of remote 3D printing will be exhibited within the main four parameters that are controlling this process: design, materiality, printing calibration/feedback, and the remote cyber-physical printing mode. Although each of these parameters with its included aspects is interdependent on the other parameters, affecting and affected by them, the categorization of the results into these four main categories simplifies the workflow for the designers and architects. The sequence of presenting the results of this study will be altered according to the sequence of the workflow that was applied in this remote, cyber/physical 3D printing, starting with material fine-tuning, followed by 3D printing calibration and reaching to the design optimization.

7.5.1 MATERIAL TUNING

The experimented clay-based materials based on varying the mixture percentages of the Ukrainian ceramic clay, marble dust, and sawdust were chosen for their availability and affordability. Besides, they are also cost-effective materials in the area where the manufacturer is located in Ukraine. From the end of the twentieth century, Ukrainian soft ceramic clay had been vastly popular in the European market with its abundance and plastic properties, which allow working on a big scale. However, there was not a considerable investigation in using this material in large-scale construction projects. Marble dust as an industrial waste increases the mechanical strength of clay and aggregates the composite environmentally friendly. The third used component, which is sawdust, is a crucial component in this mixture as its addition as a stabilizer protects adobe against moisture decomposition. Being one of the by-products from timber industries, the lignin inside sawdust has a good binding ability with clay particles in the presence of moisture and it increases the hydro-physical properties of adobe by showing improvement in the water absorption properties. Besides, using sawdust as a stabilizer reduces the dry density of the adobe. Thus, in order to achieve both the need of improving the properties of clay and also to make use of the industrial by-products, sawdust was chosen in this study as the main stabilizer.

7.5.2 3D Printing Calibration/Feedback-Decision Making Loop

Due to being a multiparameter-dependent process, the calibration phase was pivotal in this study by giving full insight into clay-based material properties, 3D printing machine properties, and the possible challenges of remote manufacturing when applying this on a limited scale. The used 3D print paste extruder was selected upon its favorable characteristics. Utilizing piston pressure (syringe-based extrusion) rather than the pneumatic (air-pressure-based extrusion) and screw-based extrusion is advantageous, allowing fabricating the high mechanical strength and viscosity materials with complex 3D structures at high resolution. On the other hand, the downside in the 3D print paste extrusion is the limited extrusion speed due to the high torque requirement. However, this limitation of 3D print paste extrusion was not encountered in this study due to the moderate viscosity of the printed materials.

As mentioned previously, successful 3D print paste extrusion depends on various parameters that are all directly and indirectly linked to each other. Although the mechanism behind the 3D print paste extrusion process can be considered fairly easy compared to the heat required or powder-based 3D print processes, due to the complex rheological properties of soft solids, the printer needs to be calibrated well. The main specifications affecting printing quality based on the printer related parameters are the nozzle diameter, the print height, the nozzle speed, and the extrusion rate. Later, to ensure a successful printing, calibration needs to be done by considering the relation between the printer and the rheological properties of the printing material sustaining a good binding mechanism. In general, until matching with the desired line diameter, the calibration parameters are altered and developed in harmony between the material and machine.

In this study, the printing mood was a non-phase-changed extrusion, in which the viscosity of the soft material is critical to be both low enough to allow extrusion and high enough to support the structure post-deposition. Therefore, not only possessing adequate mechanical strength to maintain the printed shape but also shear-thinning behavior to be easily extruded out from the nozzle are significant factors. The different three tested materials exhibited easy extrusion due to the relatively high moisture content ranging from 25 to 17 percent. Although the plasticity of clay helps in shear thinning and easy printability, it increases the water retention capacity, which causes linear shrinkage when it gets dries. During line testing, the binding mechanism, the machine, and material harmony can be considered well. On the other hand, later the high moist caused the collapse of the printed model after eight to ten layers, as well as post-drying cracking.

In the calibration with the line printing tests, the different three materials exhibited different viscosity, stability, and coherence. Figures 7.5, 7.6, 7.7, and 7.8 exhibit the calibration results with the three different material compositions.

FIGURE 7.6 Line test calibration of material composition 1: composed of 75 percent clay and 25 percent water content, the figure reveals the high liquidity of the material resulting in over-extrusion.

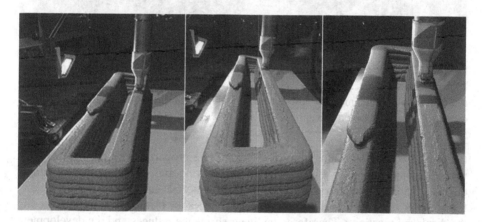

FIGURE 7.7 Line test calibration of material composition 2: composed from 55 percent clay and 27 percent marble dust, and 18 percent water. Decreasing water content and the addition of marble dust reduced the viscosity and enhanced stability. However, the marble dust did not offer a complete solution to the expansive soil properties of the clay. Besides, increasing the total weight of the composite contributed to the tendency of cracking and collapsing at the corners.

It is also obvious that the printing temperature and humidity condition affected the evaporation rate, which leads to the post-deposition spreading problem. However, it was debatable to provide an external heater or blower to accelerate the drying process, as this could cause cracking due to the sudden temperature differences at materials; see Figures 7.9 and 7.10.

On the other hand, due to the high clay content in the composite and the expansive soil properties of Ukrainian clay, cracking occurred post-drying. Because of

FIGURE 7.8 Line test calibration of material composition 3: composed from 55 percent clay, 23 percent marble dust, 2.2 percent sawdust, and 17 percent water. The addition of sawdust as a stabilizer boosted the material post-deposition stability and limited the spreading as well as enhancing the compression strength.

the greater volume of water, clay shows expansive properties, which leads to large volume changes upon drying. The danger of cracking and shrinkage due to volume changes can be minimized with coarse-grained materials such as sawdust. Applying fresh fine sawdust increases the ductility and reduces the brittle behavior by lowering the displayed plasticity. Therefore, the linear shrinkage reduces and the development of desiccation cracks decreases.

However, although composition 3 shows advancement compared to compositions 1 and 2, in printed layer accuracy, stability, and binding mechanism, the post-desiccation cracking in the lower layers started and eventually it collapsed again after the 15th layer as exhibited in Figure 7.11.

7.5.3 Design Optimization

Although the design was prepared for the printing process by using Grasshopper 3D, calibration encountered difficulty in reading the printing data (code) by the Simplify3d software used by the manufacturer. On the other hand, after solving this issue of data mismatching, the design printing data was readable by the manufacturer software. The calibration process results revealed the material spreading

FIGURE 7.9 The optimized printed design with the optimized material, composition 3, is exhibiting high printability, stability as well as a binding mechanism.

out of the path of the design, causing the deformation of the final printed form. This was due to the relatively low viscosity of the printing material composition at the beginning. Although the nozzle size was considered while designing the form, the printing conditions of low-temperature high humidity amplified the spreading coefficient. This issue was solved by altering the design form from negative (void) to positive (solid), following the same geometry and mathematical base. Figure 7.12 exhibits the linear negative initial design form and the optimized positive design form.

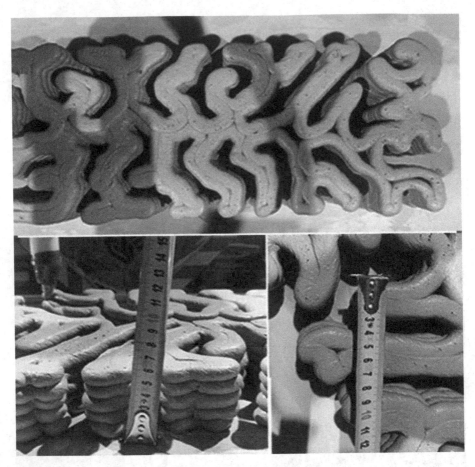

FIGURE 7.10 The final printing calibration result shows the dimensional accuracy and moderate stability of the composition 3.

7.5.4 REMOTE PRINTING

During the remote 3D printing process, many technical issues were encountered, mainly caused by the data mismatching and software incompatibility between the designer and the manufacturer, the material composition, and the printer specification incompatibility with the design. However, the accurate monitoring of the remote 3D printing process in real-time facilitated the prompt feedback-decision making loop to solve these issues. In this part, snapshot collection played a crucial role in managing the collected data from the manufacturing factory floor, and sort and store the information in an efficient fashion. Basically, reducing the required disk space and the processing time can be a solution to give an accurate insight into the printer and material performance through the production process. These snapshots are only taken once a significant change has been made to the status of the monitored process. These snapshots are also useful in the long run as by accumulating these snapshots, the

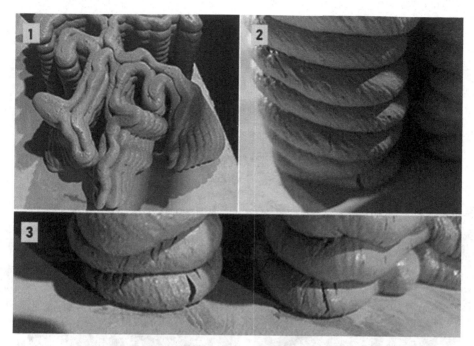

FIGURE 7.11 The post-desiccation cracks caused by the expansive soil properties of the Ukrainian ceramic clay in the material composition 3: 1, the collapse of the printed model after the partial desiccation of the lower layers; 2, and 3, the details of the cracks.

history of the process and its particular assets can be constructed and used as a guide in the future for similar manufacturing cases.

Snapshots also facilitate the similarity identification, as in cyber-physical manufacturing and due to the availability of information from several networked machines, the likelihood of capturing certain failure modes in a shorter time frame is higher. Therefore, the similarity identification section has to look back in historical records of a similar process to calculate the similarity of current machine behavior with former assets utilization.

It is also concluded from this case study that networking needs to be performed more automatically, including the equipment remote control and configuration and the intercommunication and task synchronization at least in the design phase in this particular case study. These features allow remote monitoring and automatic synchronization of the machines involved in a fabrication process with minimum human intervention and enable support for new scenarios in which any complex design can be implemented in a distributed fashion by splitting it among several networked machines (Cornetta et al., 2018). Moreover, this automation approach allows the possibility to develop Web interfaces allowing remote access to fabrication resources and thus enabling the implementation of distance-learning courses and Ed-to-Ed (i.e., Education to-Education) scenarios in which partner institutions share expensive fabrication equipment for teaching and research.

FIGURE 7.12 The design optimization based on the calibration-feedback decision making: (a) the original design (linear negative) to the left, and the optimized design (solid positive) in collaboration with the manufacturer designers, to the right; (b) the printing test results of the initial linear negative design using composition 3, exhibiting the deformation caused by the mismatching of the design specifications, with the printing conditions and material properties; (c) the printing test results of the optimized design model to the positive solid form using composition 3, exhibiting the well-defined design resolution, emphasizing the necessity of minimizing the moist content and increasing the viscosity of the printing material in case of printing linear forms without a self-supported design.

Remote 3D printing is very challenging and raises several issues related to systems networking, scalability, security, quality of service, as well as real-time and non-blocked communications. The possibility to manage collaborative and distributed fabrication processes with little or no human supervision is a powerful feature, but also a potential source of harmful failures and physical damages due to design errors that could result in fabrication equipment malfunction. This, in turn, entails the development of monitoring software and hardware infrastructure to guarantee safe equipment operation and fault tolerance (Cornetta et al., 2018). Therefore, in this study, and due to the limited scale of production, a more shared collaborative mode integrated fairly a half-automated cyber/physical manufacturing. Reducing the possible errors that result from digital data complete automation in the networked production mode and enabling more flexibility in responding to the calibration-feedback loop is based more on the designer's cognition and manipulation of the presented attributes.

7.6 CONCLUSION

This study is part of an ongoing collaborative research project on developing biocomposite materials for architectural construction applications. The presented study experiments with the cyber-physical manufacturing mode for remote 3D printing of these biocomposite materials. This was analyzed through a case study of remote 3D printing of clay-based materials in printing a coral design form biolearned from nature. By following the biological branching mathematical rules and using parametric design software Rhinoceros 3D+ Grasshopper with the integration of the physically simulated and optimized by Kangaroo and Anemone plug-in the design was generated. This parameterization of the design form generation and preparation for production was intended to facilitate the continuous modifications on the design form and specifications according to the calibration-feedback loop in the 3D printing process. The innovation of the presented work lies in the application of cyber/physical manufacturing mode on a limited resources collaboration between research institute and 3D architecture 3D printing startup. The significant aspect of this study is synthesizing three different material compositions and calibrating them remotely in the 3D printing process while maintaining real-time calibration-feedback networking and communication between the designer iBAG and the manufacturer. The calibration process included the first phase of line testing material calibration to analyze the interdependent relation between the material properties, and the printer machine specifications. The second phase included the analyses of the interdependent relationship between the material properties and the design's geometrical and mechanical compatibility. Lastly, the third phase was resulting in design optimization to fit the printer specifications. These phases revealed that the clay-based material exhibit low coherence and stability in and post-printing process causing the spreading, dimensional deformation, and post-desiccation cracking and collapse. The experimentation revealed enhanced mechanical properties of these clay-based materials by reducing the water content and the Ukrainian ceramic clay in the formula and the addition of stabilizer, which was sawdust in this study, which enhanced coherence, integration, and stability of the printed design. The experimentation also revealed the significant effect of the printing environment, especially temperature and humidity on the evaporation rate and the overall print solidification and stability. Despite the multiple negative results obtained in this study, the result was favorable in the remote 3D printing experimentation with testing the efficiency of this remote networking, communication, and forcing the critical calibration-feedback decision making. The results obtained from analyzing the efficiency of the remote cyber-physical manufacturing highlighted the importance of balancing the human–automation intervention in monitoring and managing the data collected from the production process, in order to simplify the remote manufacturing processes. It has also revealed the importance of unified design to production software between the designer and manufacture in order to avoid data-mismatching issues. Finally, it was highlighted through the case study the significant importance of snapshots that monitor the manufacturing assets and machine performance, which played a crucial role in manufacturing data collection and critical decision making.

REFERENCES

ABB. Big data and decision-making in industrial plants. http://new.abb.com/cpm/industry-software/MOM-man-ufacturing-operations-management/big-data-analytics-decision-mak-ing 2016.

A. A. Abd El Halim and A. A. El Baroudy, Influence of the Addition of Fine Sawdust on the Physical Properties of Expansive Soil in the Middle Nile Delta, Egypt, *Journal of Soil Science and Plant Nutrition,* 2014, 14 (2), 483–490.

G. Cornetta, A. Touhafi, F. J. Mateos, and G. M. Muntean, *A Cloud-based Architecture for Remote Access to Digital Fabrication Services for Education,* in 4th IEEE Cloudtech. Brussels, Belgium, November 26–28, 2018.

U. Divakarla and G. Kumari, An Overview of Cloud Computing in Distributed Systems, AIP Conference Proceedings, 2010, 1324, 184.

Foundation NS. Program solicitation: cyber-physical systems (CPS). www.nsf.gov/pubs/2015/nsf15541/nsf15541.htm 2016.

Germany Trade & Invest. Industrie 4.0: smart manufacturing for the future. 2016.

O. H. Jasim and D. Cetin, Effect of Sawdust Usage on the Shear Strength Behavior of Clayey Silt Soil, *Sigma Journal Engineering and Natural Sciences,* 2016, 34 (1), 31–41.

J. Lee, E-Manufacturing—Fundamentals, Tools, and Transformation, *Robotics and Computer-Integrated Manufacturing,* 2003, 19 (6), 501–507.

J. Lee, B. Bagheri, and C. Jin, Introduction to Cyber Manufacturing, *Manufacturing Letters,* 2016, 8, 11–15.

J. Lee, B. Bagheri, and H. Kao, A Cyber-Physical Systems Architecture for Industry 4.0-based Manufacturing Systems, *Manufacturing Letters,* 2015, 3, 18–23.

J. Lee, H. A. Kao, and S. Yang, Service Innovation and Smart Analytics for Industry 4.0 and Big Data Environment, Procedia CIRP, 2014, 16, 3–8.

J. Lee, E. Lapira, B. Bagheri, and H. Kao, Recent Advances and Trends in Predictive Manufacturing Systems in Big Data Environment, *Manufacturing Letters,* 2013, 1 (1), 38–41.

J. Lee, E. Lapira, S. Yang, and H. A. Kao, Predictive Manufacturing System Trends of Next Generation Production Systems. In: Proceedings of the 11th IFAC Workshop on Intelligent Manufacturing Systems, 2013, 150–1566.

MTConnect, www.mtconnect.org/ 2016.

K. K. Patel and S. M. Patel, Internet of Things—IoT: Definition, Characteristics, Architecture, Enabling Technologies, *Application & Future Challenges,* 2016, 6, 6112–6131.

Y. Riahi and S. Riahi, Big Data and Big Data Analytics: Concepts, Types and Technologies, *International Journal of Research and Engineering,* 2018, 5(9), 524–528.

P. S. Seemma, S. Nandhini, and M.Sowmiya, Overview of Cyber Security, *International Journal of Advanced Research in Computer and Communication Engineering,* 2018, 7 (11), 125–128.

J. Shi, J. Wan, H. Yan, and H. Suo, A Survey of Cyber-Physical Systems. In: International Conference on Wireless Communications and Signal Processing (WCP), 2011, 1–6.

8 Employing *Columba livia* Swarmal Patterns in Designing Self-Sufficient Photo Bioreactor of *Chlorella* spp Cultivation in Plaça de Catalunya

Javier G. Castillo, Andrew Gennett,
Alberto T. Estévez, and Yomna K. Abdallah

8.1 INTRODUCTION

Our cities are complex artificial ecosystems, in which we have sought the sterilization of the natural world in our built environment. However, these sterilized isolated architectural capsules have proved their failure to maintain human being's physical, mental, and emotional health exemplified during the current critical pandemic. The mass-scale lockdown of cities, trapping humans away from any interaction with the natural ecosystem, proved that the "biophilia" shouldn't be considered a luxurious tendency that some beings enjoy more than the others, but rather a physiological

DOI: 10.1201/9781003240129-8

need that maintains human well-being. Humans are designed to live and interact with nature on a daily basis. Thus, our architectural design of today and future cities must shift toward supporting biodiversity, symbiotic relations, boosting social–natural interactions, and maintaining the ecosystem. Emerging from this biophilic approach, in direct contrast to the isolated architectural islands mode, this work aims to present a role model of multi-interaction between different species serving each other to maintain biodiversity in the built environment ecosystem. The inclusive model is designed to increase the biodiversity of city centers while closing the production cycles and constricting the production to proximity markets. This is to be achieved through the study of *Columba livia* flocking (swarm behavior) and its constraints in the open urban areas (plazas), and how this affects the social flocking patterns of the locals in Barcelona's Plaza de Catalonia. *C. livia* is a lovable typical species of pigeons that inhabit the Iberian Peninsula (CABI Datasheet: *C. livia*, pigeons, 2019), these friendly birds are adored by the locals which is reflected in their behavior of feeding these birds, especially in open urban areas, parks as well as plazas. This direct relationship between both flocking patterns, of birds and humans are essentially interwoven as one. At this point, our third essential element in this symbiotic relationship between humans and *C. livia*, is the "food" that is used to build this relation influencing this behavioral pattern. Moreover, algae as an element is even more interesting as it has its own unique growth pattern that can be employed in function and form of architectural and urban design. Additionally, the microalgae *Chlorella* spp have an immense nutrition value for birds.

In this work a detailed study of this tertiary symbiosis between *C. livia*, *Homo sapiens*, and *Chlorella* spp is analyzed through the simulation of the behavioral patterns of these three symbionts via a mathematical model that was used in a form-finding process for the design of the architectural and urban built environment, while proposing this interactive behavioral simulation data as real-time data visualization. The resulting form is scaled to produce an urban product to achieve social well-being, biodiversity, efficient product design, nature environment maintenance, and economic revenue, as exhibited in Figure 8.1.

8.2   ANALYZING THE TRI-SYMBIOSIS: *COLUMBA LIVIA, HOMO SAPIENS,* AND *CHLORELLA* SPP

The mutualistic relationship between people and pigeons have gone through many reforms since the cities began to expand vertically with high-rise buildings ("Rock Pigeon." *All About Birds*). From a source of food to a key component of our communications network, and as an inspiration to Charles Darwin for the concept of Natural Selection (Darwin and Kebler 1859), yet from such humble beginnings how have they fallen to be considered as pests by the majority! But not to all for some, these creatures offer companionship that compensates the ever-growing isolation in the modern cities (Cox and Gaston 2016).

In many cities around the world, you will find a site where the birds flock following the pattern of people that are feeding them bits of bread and grain. The joy is most prevalent with the children who play among the flocks, and these interactions through

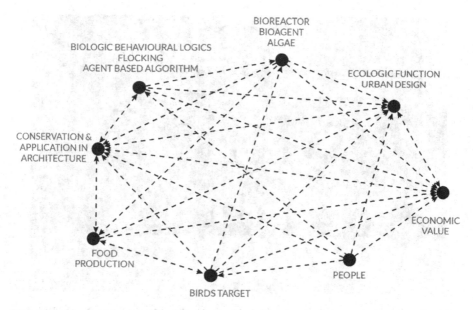

FIGURE 8.1 Diagram of the integrated urban design solution based on the behavioral, functional, and formal analyses of the tri-symbiotic relation between *Columba livia*, *Homo sapiens*, and *Chlorella* spp (diagram by authors).

feeding have a profound effect as seen by their smiles and heard by their laughter. However, these actions have a negative effect on both—the birds and the people, so a proper balance is critical to cities and wildlife management agencies such as the Wildlife Ecology Service (SEFaS).

One of the main drawbacks of these friendly creatures is the economic damage that they cause, in two main ways: first by ruining the architectural and urban built environment by the accumulation of their excrement, as in the damaging effect of *C. livia* excrement accumulating on architectural and urban elements in the built environment, causing hazardous effects on human health.

C. livia typically reside in rock formations, settling in crevices to nest. They nest communally, often forming large colonies of many hundreds of individuals (Audubon Society). The health hazards to humans are demonstrated by the severe acute respiratory syndrome (SARS) and bird flu over the years, of which pigeons act as vectors in the zoonotic disease infecting humans (Mansour et al. 2014). The second economic damaging effect of these birds is the profound budgets that some European governments spend over feeding these birds as well as the cleaning costs associated with the excrement removal. In Barcelona, the case study in the current work, the municipality spends around €250,000 per year for feeding the pigeons in an effort to control their populations (Lyne 2017). This is conducted through the automated feeders that release maize kernels (Figure 8.2), an automated feeder coated in the contraceptive (Nicarbazin), which affects development of the egg yolk (Lyne 2017).

FIGURE 8.2 Automated feed dispenser. Image Credit: Massimiliano Minocri/*El País* (Lyne 2017).

Thus, an integrated solution that aims to lessens the economic drawbacks while maintaining the healthy social relation between people and birds, and in respect to an improved built environments ecosystem needs to be well analyzed.

On the other hand, the commensalism between humans and algae has always been beneficial for both. Algae, which is a phototroph, with environmental benefits via photosynthesis, is a perfect agent in this current project. Therefore, the proposed bioreactor, installed at the point source of the greenhouse gas generated within our cities, proves to be an optimal environmental solution to reverse the climate change. A bioreactor is a system designed for the cultivation of an organism, through maintaining its optimal environmental conditions for growth and reproduction (www.merriam-webster.com/dictionary/bioreactor.6/4/2021).

Furthermore, algae can also be consumable by *C. livia*, as rock doves prefer plant matter in their dietary system. Being rich in vitamins and proteins, it is a sufficient alternative for other conventional grains that are normally used to feed the birds.

From this analysis, it is quite clear that algae are the protagonist in this case study, and the solution lies in the mass cultivation of this bioagent, in a way that bring it to the encounter of human everyday life and activities. This integration of algae bioreactor as part of the urban and architectural built environment proposes a problem-solving method through a multiscale solution as a conceptual shift from consumption-based built environment to urban-integrated biofactories, which are intended to educate the main consumption centers and convert them into production centers, vanishing the barriers between production and leisure activities from the architectural and urban design point of view, and as a social behavior as well.

8.3 DESIGNING THE ECOLOGICAL FUNCTION

Employing algae to solve the symbiosis between humans, birds, and algae as an integrated solution for design balances the built environment ecosystem. The design solution rests upon algae as an alternative to the maize kernels used in Barcelona city to control the common pigeon populations. This solution employs a zero-kilometer alternative product grown in the same Plaça de Catalunya where it is to be used. This product is a pelletized microalga *Chlorella* spp, grown in the existing fountains in La Deessa utilizing a modified open pond photobioreactor system (Daliry et al. 2017). Replacing the conventional photobioreactor paddle wheel mixing system with a visually stimulating vertical bio-receptive surface. This surface will be in direct link to visualize the people–pigeon mutualistic relationship meta data extracted by an agent-based algorithm.

This solution would flip a cost center of the city into a self-sustaining system without increasing the workload. Additionally, an interactive harvesting component will be proposed as well as an informative tablet to involve the community in the solution for the benefit of the pigeons and people's health equally.

8.3.1 BIO AGENT: *CHLORELLA* SPP CULTIVATION

Chlorella is a eukaryotic marine *Trebouxiophyceae* strain that has large-scale commercial cultivation in Asia as a protein-rich food, a nutritional supplement, and biofuel source. It can be cultivated autotrophically, mixotrophically or heterotrophically. *Chlorella* has a quite robust growth for cultivation in open ponds as well as photobioreactors (Daliry et al. 2017). *Chlorella* originally is a seawater blue-green cosmopolitan species with small globular cells (about 2–10 mm diameter). However, some strains of *Chlorella* live in both aquatic and terrestrial habitats. It includes strains with a high temperature tolerance since some strains can grow between 15°C and 40°C.

In the current work, *Chlorella* spp was selected for culturing as it is a genus of unicellular freshwater microalgae that are fit for human consumption and are used as additives with high nutritional value in feed for agriculturally important animals. As well as that *Chlorella* is characterized by its simple cultivation, high productivity and levels of protein and other nutrients (Abdelnour et al. 2019). The *Chlorella* spp used in this study was kindly provided by BioBabes (Biocentric Design Group, UK) as a starter liquid culture kit of 250 mL. The growth medium was prepared in 1 L of distilled water that was distributed evenly afterward in two flasks, to maintain a ready-made growth medium for the following growth phases. The growth medium utilized sodium nitrate of 1.5 g/L dissolved in 1 L of distilled water (Starr 1978). After that, the medium was autoclaved for 90 min and left to cool down in a water bath until reaching 28°C. Then it was split evenly under sterilized conditions in two 500 mL flasks. A 250 mL *Chlorella* spp of 14 days grown liquid culture were used to inoculate the 500 mL medium under sterilized conditions. Then, an oxygenation device was designed of an air pump of 60 L/h pumping rate and a silicon tube. The oxygenation device was then fixed to the flasks to maintain sterilized conditions, and the flasks were incubated in 28°C (Daliry 2017) in direct sunlight for 14 days. The cultivation

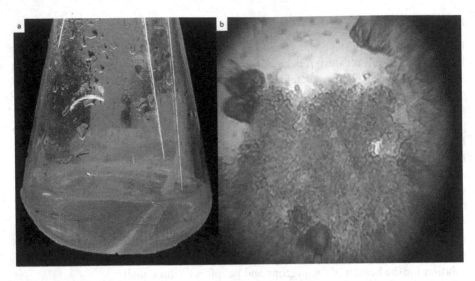

FIGURE 8.3 Cultivation of *Chlorella* spp (A) and the morphological identification under compound light microscope 20× (B) (photos by authors).

experiment was conducted in triplicate in the facilities of Universitat Internacional de Catalunya, Faculty of Health Sciences. Figure 8.3 shows the cultivation process and morphological identification under compound light microscope.

8.3.2 PHOTO BIOREACTOR DESIGN FOR MASS CULTIVATION OF *CHLORELLA* SPP

In the current study, our main objective as stated before is the wide integration of a self-sufficient system that exhibits the symbiosis relation between human/algae and birds in order to boost biodiversity and encourage the mainstream to shift their everyday social habits to environmentally friendly routine. For achieving this, this work is exhibited in a simple, easy to understand and to apply protocol, more than being in the form of a basic science research; we aim for spreading the integration of biodigital self-sufficient systems and democratizing the experience to attain environmental and economic benefit. Thus, the sophistication of the scientific process of designing and building a bioreactor hereby is broken down to simple steps and components rather than materials and methods.

The photobioreactor was designed to build and operate an open pond system, as it was decidedly the more cost-efficient production method due to simplicity, low operational costs, as well as maintenance (Costa and Morais 2014). Moreover, open pond systems are commonly preferred way around the world to cultivate *Chlorella*, since it proves to be large-scale production method. While there are disadvantages of an open pond, such as poor light utilization through the cell, poor evaporative loses, poor diffusion of CO_2 to the atmosphere, and requiring large surface area (Gupta and Singh 2018), in the current work the authors opted to break the stereotypes, manipulating the typical design of open pond system through replacing the horizontal cultivation method with a vertical surface. The vertical system increased the surface area, which allowed better light distribution to all cells, combined with a variable pumping system to control flow over the surface. Varying the water flow rate over the surface allows for control over temperature and aeration. Added to this, integrating the form-finding results of the swarm behavior simulation in the functional design of the photo bioreactor, to exploit the swarmal behavior rules of separation, alignment, and cohesion in solving the specific disadvantages of a typical open pond system (OPS) photobioreactor. The separation achieved through the different swarm patterns presented an integrated solution for the disadvantages of OPS of poor light utilization through the cell, poor evaporative losses, and poor diffusion of CO_2 to the atmosphere, while the alignment and cohesion work collaboratively with separation to increase the surface area for the attachment of *Chlorella* cells solving the need of great surface area that is required in a conventional OPS through maximizing aggregated niches for the microalgae attachment and cultivation while maintaining oxygenation, stirring, and light penetration. Added to the formal efficiency of this photobioreactor design, the bioagent as well was adequate for urban and architectural open area applications, as the *Chlorella* species can tolerate higher temperature, which make them resistant against contamination (Serra-Maia et al. 2016). And as it's said that the solution emerges from the challenge; for solving evaporation rate needed and large land required, searching the urban environment has given the solution for this, by integrating the photobioreactor in what was already built and an essential part of the urban tissue of Plaça de Catalunya. The existing fountain on one side of the square welcomes the crowds to gather around it in sunny days "para tomar el sol" and enjoy feeding the pigeons while they enjoy other social activities. This fountain exactly proposed itself as a shelter for our photobioreactor design; it was nearly asking for it.

Moreover, further optimization was conducted on the photobioreactor design replacing the typical paddle wheel that is utilized for the culture continuous mixing with a waterfall design, which further boosts the aeration of the culture. Additionally, this surface may act as a thermal control unit, by controlling the flow rate of water over the surface.

8.3.3 BEHAVIORAL MATHEMATICAL MODEL: SWARM (FLOCKING) BEHAVIORAL SIMULATION

In order to link the urban and architectural design with the symbiosis between humans and birds, the photobioreactor vertical plane was designed by using an agent-based algorithm to simulate the flocking behavior of the birds as they interact with people in Plaça de Catalunya, thus affecting the behavioral patterns of the people as well. This agent-based swarm simulation is controlled by three main parameters: separation, alignment, and cohesion of the flock in response to the stimulus "food," which is, in this case the microalgae *Chlorella* spp. Tracking these flocking movements over time as exhibited in Figure 8.4 and digitally simulating it with the agent-based algorithm produced a behavioral pattern (Figure 8.5) by which a few simple rules were applied, to construct the vertical planes interstices. Figure 8.6 shows the bio-receptive

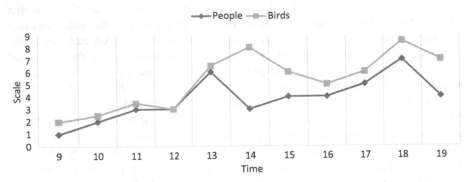

FIGURE 8.4 *Columba livia* and people swarmal behavior and interaction activity. In "dark" the line shows the number of people feeding birds in Plaça de Catalunya with a factor of 2, for every increment under people there are actually 2. In "light" the line shows the number of *Columba livia* birds with a factor of 10, meaning that for every point there are 10 birds (diagram by authors).

surface design stages from the swarmal agent-based algorithm that was used as a form-finding method. The resulting pattern based on biological behaviors is visually appealing, aiming to capture the visual attention of people in Plaça de Catalunya.

Achieving the visual attraction wasn't the only aim of this multilayered design, the auditory senses of the square visitors was also targeted by the soft white noise of a waterfall produced by the water flowing over the bioreactor's surface. This white noise will be a welcome addition over that of the local traffic in what is already a gathering point in the city. By improving the audiovisual environment of this gathering point, an expected increase in the foot traffic in this area will give the Plaza visitors a chance to engage in this interaction center.

Added to all these stimuli to engage the public in this environmentally responsible activity, an educational aspect is achieved by further information on the artwork, which is the microalgae photobioreactor, as well as information on its role in the urban ecology based on the people–pigeon–city interaction network. Figure 8.7 exhibits ecological interaction network. This information will be provided through live real-time projections on the bioreceptive surface itself during the evening hours. By providing this information to the public, they will be encouraged to participate in this environmental economic activity by purchasing the harvested *Chlorella* in the form of dried and processed pellets supplemented with the contraceptive elements to feed the pigeons.

To further optimize the microalgae *Chlorella* spp growth and production, and to boost its visual attraction, an embedded light array was used to extend the growth time with supplemental light in addition to the sun. Figure 8.8 exhibits the monitoring of the growth of *Chlorella* spp in the photobioreactor fountain as a function of turbidity of water. This constant light solution will have a dual effect, not only will it be beneficial to the microalgae growth but will further attract the public's interest due

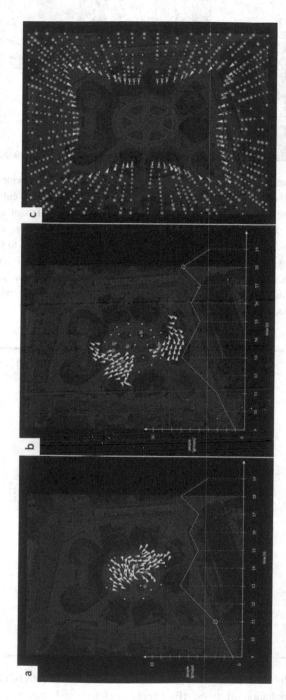

FIGURE 8.5 Main key frames from the sequence of the flocking behavior simulation of *Columba livia* agents (light), followed by human agents (dark), through different times during the day: (a) flocking behavior at 11:00 a.m., (b) flocking behavior between 6:00 and 7:00 p.m. showing that the increase in *C. livia* agents is related to the increase of people numbers in Plaça de Catalunya derived by the activity of people feeding the pigeons, (c) combined simulation of *C. livia* agents and people agents approaching the plaza from the surrounding streets (images by authors).

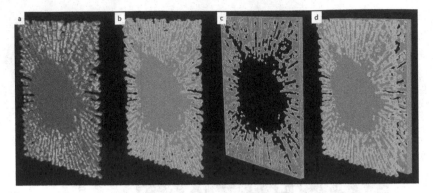

FIGURE 8.6 Stages of form-finding process extracted from the swarm behavior agent-based algorithm to design the bio-receptive surface of microalgae *Chlorella* spp cultivation photobioreactor: (a) combined swarm simulation keyframes, (b) extracting planner surfaces from the generated form, (c) level 2 of design complexity applied by subtraction, creating textured surface of solids and voids to achieve more light and oxygen penetration to the bio-receptive surface, enhancing the microalgae culture growth conditions to boost microalgae production, (d) level 3 of design complexity applied by transition and addition of the same subtracted part of the designed planner surface, adding more surface area through providing more texture and niches to increase the microalgae culture (images by authors).

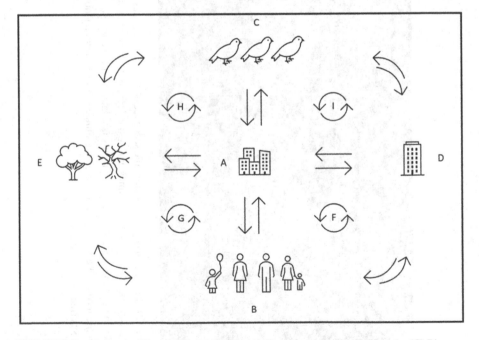

FIGURE 8.7 Ecological Interaction Network. (A) The urban setting. (B) People. (C) Pigeons. (D) Architecture. (E) Defoliation of trees. (F) People–Architecture–Urban relationship. (G) People–Trees–Urban relationship. (H) Pigeon–Tree–Urban Relationship. (I) Pigeon–Architecture–Urban relationship (diagram by authors).

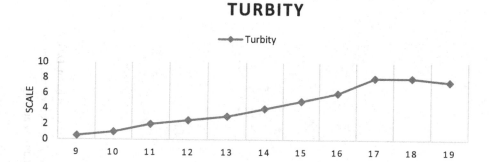

FIGURE 8.8 Monitoring of the growth in the PBR fountain as a function of turbidity of water (diagram by authors).

to the natural positively phototrophic behavior of people. Moreover, this light array could be used to create artistic patterns and utilize multicolored diodes increasing its Instagram-ability.

8.3.4 Prototyping and Digital Fabrication

Subtractive digital fabrication method was designated to fabricate the optimized design of the photo bioreactor on a pilot scale in order to be utilized in extensive study of its sufficiency in the microalgae *Chlorella* production before mass scaling it. The complex bio-receptive surface providing the shelter for the microalgae was cut by laser cutter into three sheets of acrylic that are composing the complex textured bio-receptive surface (Figure 8.9) to produce positive and negative of the swarm design pattern. These were 60 × 100 cm acrylic sheet, 5 mm thick, and 60 × 50 cm acrylic sheet, 2 mm thick. Then the cut sheets with their positive, negative, and particle parts were glued with a waterproof acrylic glue.

At this point, the authors realized that using acrylic glue isn't the perfect environmentally friendly production practice, as well as its incompatibility to be used in larger scale, for solving this, the authors are currently working on two approaches to assemble the textured bio-receptive surface parts without using such glue. These are developing a joinery system in order to attach each particle, positive and negative parts together and this of course implies hard work on fabrication files in order to produce this very miniature joinery system, the other approach is not as much environmentally friendly as the joinery system; it depends on a special technique in glass fusing to combine totally two glass or acrylic parts together by heat.

After assembling the bio-receptive surface as shown in Figure 8.10, a water pump was used in order to left the water flow from the enhanced open pond system's basin to flow over the bio-receptive surface providing equalized distribution of water, oxygen and nutrients to all *Chlorella* cells almost equally as shown in Figure 8.11.

Finally, after assembling the essential parts of the OPS photobioreactor for *Chlorella* spp cultivation, there are essential probes for monitoring and maintaining

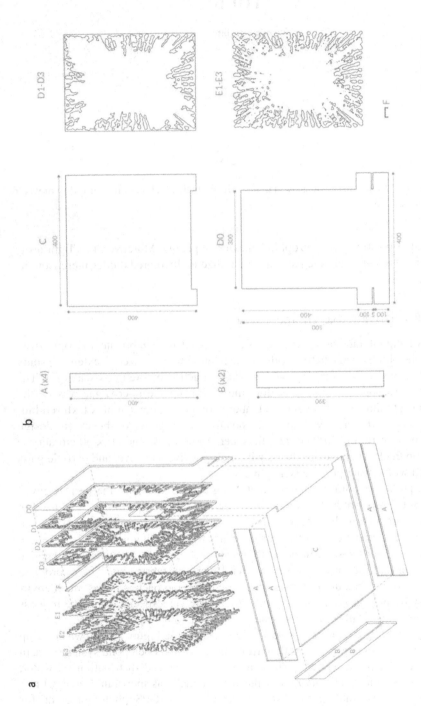

FIGURE 8.9 Schematic of the photobioreactor. (a) Exploded Components Diagram of Bioreactor. A–C constitute the water basin. D0 is a dual component completing the water basin as well as the vertical surfaces backboard. D–F slices build the vertical surface where the algae to be flow across for aeration. (b) Elevations showing the exact dimension of the scaled photo bioreactor (diagrams by authors).

FIGURE 8.10 The digitally fabricated and assembled bio-receptive surface of the OPS photo bioreactor, from the form-finding process based on swarmal behavior simulation of *Columba livia* pigeons and interacting crowds in Plaça de Catalunya (photos by authors).

FIGURE 8.11 The completed prototype of the OPS photobioreactor (photos by authors).

FIGURE 8.12 Rendering of the photobioreactor OPS implementation in urban design of Plaça de Catalunya, achieving symbiosis between *Columba livia*, *Homo sapiens*, and *Chlorella* spp (image by authors).

the performance of this systems; these are a microcontroller, a thermometer, a CO_2 sensing device, a turbidity monitoring device, and a harvesting device that is in the simplest ways can be a microfilter for separating the *Chlorella* culture from the water. These devices are still under development and will be exhibited in future works; however, it was mandatory to mention them to provide a full insight about this case study giving a simpler understanding of the proposed methodology of self-sufficient urbanism in support of biodiversity and ecosystem interaction. Another futuristic aspect is including multimedia as a visual effector and educator in the adoption of environmentally friendly practice among the folks on an urban scale, this was through real-time projections of data visualization mapping the flocking behavior of *C. livia* agents during the day and combining these patterns into artistic multimedia projections on the bio-receptive surface. Figure 8.12 shows a visualization of the urban implementation of the photo bioreactor OPS in Plaça de Catalunya.

8.4 CONCLUSION

In the presented work, an integrated solution was proposed for boosting environmentally friendly practices on everyday life bases, as well as encouraging participation in maintaining biodiversity and healthy interactions between the urban design and the ecosystem. This solution was based on a tri-symbiosis relation between *C. livia*, *Homo sapiens*, and *Chlorella* spp, as the main partners involved in this solution action. Exploiting natural social tendency of humans to interact with pigeons through feeding them, and the resulting pattern extracted from the swarm behavior that these pigeons perform in open urban area; the design solution proposed the microalgae *Chlorella* spp as the connecting loop in this process, being a favorable dietary element to both humans and pigeons, as well as being cost-effective, environmentally friendly organism that consumes CO_2 and generates oxygen and nutrition. Thus, the behavioral swarmal pattern was employed in a photobioreactor design of open pond system optimized design for the mass cultivation of the *Chlorella* spp microalgae that will be after processed as a pelleted food for the pigeons in open urban areas, from this multiscale solution to the interconnected net of interdependencies were solved following an environmentally friendly approach and achieving a closed product cycle.

REFERENCES

S. A. Abdelnour, M. E. Abd el-hack, M. Arif, A. F. Khafaga, and A. E. Taha, The application of the microalgae *Chlorella spp.* as a supplement in broiler feed. Published online by Cambridge University Press, April 5, 2019.

J. A. V. Costa, M. G. Morais, Chapter 1—An Open Pond System for Microalgal Cultivation. *Biofuels from Algae*, 2014, 1–22.

D. T. C. Cox and K. J. Gaston, Urban Bird Feeding: Connecting People with Nature. *PLoS One*, 2016, 11(7), e0158717.

S. Daliry, A. Hallajsani, J. Mohammadi Roshandeh, H. Nouri, and A. Golzary, Investigation of optimal condition for *Chlorella vulgaris* microalgae growth. *Global Journal of Environmental Science and Management*, 2017, 113(2): 217–230.

C. Darwin and K. Leonard, *On the Origin of Species by Means of Natural Selection, or, the Preservation of Favored Races in the Struggle for Life.* London: J. Murray. 1859.

"Datasheet: *Columba Livia* (Pigeons)." CABI Datasheet on *Columba livia* (pigeons), November 19, 2019. www.cabi.org/isc/datasheet/87913#.

N. Gupta and D. P. Singh, Low-cost Production of Algal Biofuel from Wastewater and Technological Limitations, *Emerging Energy Alternative for Sustainable Environment*, The Energy and Resource Institute, August 2018.

"Help us control pigeon overpopulation don't feed them!" Ajuntament Barcelona.cat. November 16, 2018. https://ajuntament.barcelona.cat/ecologiaurbana/en/noticia/help-us-control-pigeon-overpopulation-dont-feed-them_736715.

F. Liang, X. Wen, Y. Geng, Z. Ouyang, L. Luo, and Y. Li, Growth rate and biomass productivity of *Chlorella* as affected by culture depth and cell density in an open circular photobioreactor. *Journal of Microbiology and Biotechnology* 2013, 23(4), 539–544.

N. Lyne, Barcelona to feed pigeons contraceptives in bid to slash numbers. February 22, 2017. https://english.elpais.com/elpais/2017/02/22/inenglish/1487760689_433843.html.

S. M. Mansour, R. M. El Bakrey, H. Ali, D. E. Knudsen, and A. A. Eid, Natural infection with highly pathogenic avian influenza virus H5N1 in domestic pigeons (*Columba livia*) in Egypt. *Avian Pathology* 2014, 43(4), 319–324.

"Rock Pigeon." *All About Birds*. Cornell Laboratory of Ornithology. Accessed February 19, 2008.

"Rock Pigeon *Columba livia*." Audubon Society Webpage. Accessed March 20, 2021.

R. Serra-Maia, O. Bernard, A. Gonçalves, S. Bensalem, and F. Lopes, Influence of temperature on *Chlorella vulgaris* growth and mortality rates in a photobioreactor. *Algal Research*, 2016, 18, 352–359.

R. C. Starr. The culture collection of algae at the University of Texas at Austin. *Journal of Phycology*, 1978, 14 (Suppl. 4), 47–100.

9 Addressing Agricultural Pressures on Water Resources

A DEA Environmental Assessment in the Case of European Transboundary Basins

A. Expósito and F. Velasco

9.1 INTRODUCTION

Agriculture is not only the main user of water resources (around 70 percent of all water abstraction worldwide) (FAO 2017) but is also the most intractable source of riverine and marine pollution (Boyle 2014; Mohaupt et al. 2007). Most European river basins (RBs) face significant pressures from agriculture, both in terms of quantity (e.g., abstraction stress) and quality (e.g., agrochemical diffuse pollution) (EEA 2015; EC 2012; Mohaupt et al. 2007; Özerol et al. 2012). The European Union (EU) water legislation aims to achieve an ecologically and chemically good status of water bodies through the reduction of demand pressures on water resources, a progressive reduction of pollution, and the preservation of ecosystems (van Rijswick et al. 2010). The European Environment Agency (EEA) (EEA 2012a, 2015) indicates that the environmental status of a large number of water bodies across Europe is unacceptable and is of greater concern in large transboundary river basins (TRBs). Furthermore, the EEA asserts that emissions of nutrient (e.g., phosphorus and nitrogen) compounds from agriculture represent a serious pollutant threat to European water bodies (EEA

DOI: 10.1201/9781003240129-9

2012b; EU 2017). The EU Water Framework Directive (WFD) (EC/2000/60) explicitly includes transboundary rivers, for whose management international or interregional structures are envisaged (Art. 3, Preamble 35). However, this directive fails to specify instruments for coordination and responsibilities for riparian states so that the established environmental goals may be more easily achieved (van Rijswick et al. 2010; Wiering et al. 2010).

Transboundary rivers represent complex interdependent structures at a multidimensional level (e.g., socioeconomic, political, environmental) for riparian countries, whereby the river is a source of externalities (e.g., agrochemical pollution) and mutual vulnerabilities (Dimitriou et al. 2012; Dinar et al. 2013; Kauffman 2015). In this respect, the subtractability and nonexclusion characteristics of transboundary rivers have led them to become common pool resources shared by riparian countries where the common dilemma of overuse and mismanagement can easily occur (Ostrom 2005). Much of the recent research on the management of transboundary rivers has focused on the determinants of political and geographical conditions for effective cooperative management (e.g., Dinar et al. 2013; Zawahri and Mitchell 2011), legal aspects and conflict resolution (e.g., Jager 2016; Petersen-Perlman et al. 2017), and risks associated to climate change, such as droughts and floods (e.g., Bakker and Duncan 2017; Pulwarty and Maia 2015). Although environmental degradation is considered an incentive for international joint efforts (Dinar et al. 2013) and agricultural pressures on water resources represent a significant challenge for RB management (Kallioras et al. 2006), their analysis has attracted much less attention in academic and international cooperation spheres (Munia et al. 2016; UNECE 2011).

The present work aims to fill this gap and offers a comparative analysis for the identification of factors that may influence performance patterns in managing agricultural pressures. With this aim, the data envelopment analysis (DEA) methodology used in this study enables the assessment of the capacity of the analyzed RBs regarding the minimization of agricultural pressures on water resources.

The rest of the paper is organized as follows. Section 9.2 briefly describes our case study comprised of 20 European RBs. The following section (Section 9.3) presents the DEA methodology used and describes the data. Section 9.4 summarizes the results of our analysis. Finally, certain concluding remarks are offered in Section 9.5.

9.2   CASE STUDY

TRBs comprise about 47 percent of the world's continental land area and Europe has the largest number of these transnational rivers, that is, 68 out of the total 286 in the world (Giordano et al. 2014). The watershed or RB scale has often been considered a useful organizational unit for the evaluation of policy and research issues, especially regarding environmental concerns (EC 2012). Further, the WFD recognizes the RB as the main natural unit for the protection of the status of water bodies, and as the appropriate scale for integrated water resource management (IWRM) (Berbel and Expósito 2018). In this respect, the DEA method applied herein, which is explained in greater detail in the subsequent section, uses RBs as the decision units to be assessed. Specifically, the method applied in this study allows us to assess the management performance of these decision units (or RBs in our case study) at reducing agricultural pressures, such as nutrient pollution and abstraction stress, on water resources.

Over recent decades, cooperation efforts, involving all TRBs to solve environmental problems derived from human activities (including agriculture), have increased dramatically, as globally initiated by the Global Environment Facility (GEF) set up by the World Bank in 1991 (Gerlak 2004), and followed by the Convention on the Protection and Use of Transboundary Watercourses and International Lakes (UNECE, Helsinki, 1992; Libert 2015). To this end, the Convention established the principle of the joint management bodies, based on any multilateral institutional arrangement for cooperation between the riparian countries. Since then, most European TRBs have created international commissions to jointly manage water resources on a basin scale and to reduce negative externalities from human activities, such as agricultural pollution and abstraction stress. One of the first multilateral cooperation institutions was established in 1988 through the International Commission for the Protection of the Danube river (ICPDR), whose main objective is the implementation of the Danube River Protection Convention. Similarly, international commissions have been established for many other rivers in Europe, such as the Conventions of the International Commission for the Protection of the Elbe (1991), the Oder (1999), the Rhine (1999), and the Albufeira Convention for the Iberian rivers (1998). As further discussed in the "Discussion" section of this study, these transboundary commissions have been widely criticized for failing to have achieved effective joint management aimed at reducing pressures on water resources (including those of the agricultural sector), and hence for failing to have served as an effective instrument to improve the ecological and chemical status of TRBs (Bernauer and Kuhn 2010).

In this context, this study takes a representative sample of 20 European RBs to assess their management capability in terms of reducing agricultural pressures on water resources. It is worth noting that the area of some RBs, such as Ebro and Po, are mainly located in one country. This study aims to assess the managerial efficiency at RB scale, thus taking the RB as a management decision-making unit, regardless of the number of riparian countries or the existence of TRB agency. Nevertheless, the potential effect of these factors, among others, on the estimated efficiency patterns are also analyzed in this study.

Table 9.1 shows various characteristics regarding drainage area, population, per capita gross domestic product (GDP), renewable water sources (RWS), total water withdrawals, and the relative weight of agriculture on those total withdrawals. These selected indicators show the high heterogeneity in our sample of RBs. A further description of the riparian countries of each TRB is given in Table 9.2.

9.3 METHODOLOGY AND DATA

9.3.1 METHOD

The DEA method was initially proposed by Farrell (1957). Subsequently, it has been extended under various functional schemes, such as an input-oriented scheme with constant returns to scale (Charnes et al. 1978), output-oriented maximization (Charnes et al. 1981), variable returns to scale (Banker et al. 1984), and both radial

TABLE 9.1
Indicators of Selected European TRBs (2010)

River basin	Drainage area (10^3 km²)	Population (10^3)	GDP pc (USD)	RWS (km³/ year)	Water withdrawals (km³/year)	Agric. withdrawals (% total)
Danube	796	80,185	18,478	221,762	53,822	36.0
Dnieper	511	29,457	5,889	66,635	14,486	35.5
Don	439	18,819	11,359	45,375	10,205	29.4
Douro/Duero	97	3,492	25,521	24,098	7,412	79.5
Ebro	85	2,805	28,024	19,082	9,865	72.3
Elbe	139	21,860	37,940	28,957	7,462	9.8
Guadiana	67	1,475	28,017	11,076	8,605	94.0
Kemi	54	105	44,504	17,900	30	2.7
Klarälven	50	901	60,573	20,564	602	12.7
Maritsa	53	3,476	8,965	11,970	6,404	52.6
Neman	93	4,789	12,144	20,754	880	12.1
Oder/Odra	119	15,718	15,163	20,997	4,720	4.1
Po	72	15,918	35,500	48,958	18,575	40.5
Rhine	164	48,831	49,543	74,972	28,834	4.5
Rhone	97	10,055	46,047	52,339	8,207	25.1
Seine	73	15,775	41,422	20,712	8,353	20.6
Tagus/Tejo	71	7,244	28,303	19,297	7,968	59.9
Vistula/Wista	192	23,148	12,753	34,604	7,699	3.4
Volga	1,412	58,621	14,612	274,165	25,004	11.4
Vuoksi	287	3,246	23,001	87,344	5,590	1.3

Source: Data from TWAP (2019).

and nonradial approaches (Sueyoshi and Sekitani 2009), among others. In the water management sector, DEA methods have been extensively used to assess efficiency among a group of management or decision-making units (DMUs) (e.g., water utilities) (Xiang et al. 2016; Romano et al. 2017, among others). Zhu (2016) offers a recent review of DEA literature and applications in environmental issues.

DEA methods are based on a nonparametric approach to estimate efficient frontiers in the sense that no assumption regarding the functional form is required. This enables relative efficiency estimates to be obtained for a group of DMUs (i.e., our sample of European TRBs) by using multiple inputs and outputs and alternative output–input specifications. Furthermore, DEA methods measure the efficiency of a DMU with the simple restriction that all sampled DMUs lie on or below the efficient frontier and obviate the need to assign prespecified weights to either inputs or outputs. Each DMU not on the frontier (thus, an inefficient DMU) is scaled against a convex combination of the DMUs on the frontier faced closest to it. Thus, efficiency mappings of a group of DMUs can be obtained and efficient (or benchmark) DMUs can be identified. Additionally, and conversely to other methodologies, such as qualitative analysis and multicriteria schemes, DEA methods are capable of identifying

TABLE 9.2
Riparian Countries

TRB	Riparian countries
Danube	Albania, Austria, Bosnia And Herzegovina, Bulgaria, Croatia, Czech Republic, Germany, Hungary, Italy, (The former Yugoslav Republic of) Macedonia, (Republic of) Moldova, Montenegro, Poland, Romania, Serbia, Slovakia, Slovenia, Switzerland, Ukraine
Dnieper	Belarus, Russian Federation, Ukraine
Don	Russian Federation, Ukraine
Douro/Duero	Portugal, Spain
Ebro	Andorra, France, Spain
Elbe	Austria, Czech Republic, Germany, Poland
Guadiana	Portugal, Spain
Kemi	Finland, Norway, Russian Federation
Klarälven	Norway, Sweden
Maritsa	Bulgaria, Greece, Turkey
Neman	Belarus, Latvia, Lithuania, Poland, Russian Federation
Oder/Odra	Czech Republic, Germany, Poland, Slovakia
Po	France, Italy, Switzerland
Rhine	Austria, Belgium, France, Germany, Italy, Liechtenstein, Luxembourg, Netherlands, Switzerland
Rhone	France, Italy, Switzerland
Seine	Belgium, France
Tagus/Tejo	Portugal, Spain
Vistula/Wista	Belarus, Czech Republic, Poland, Slovakia, Ukraine
Volga	Kazakhstan, Russian Federation
Vuoksi	Belarus, Finland, Russian Federation

input–output relationships that remain unobserved for other methods. DEA methods therefore constitute a powerful instrument in obtaining additional strategic information not provided by alternative methodologies.

DEA methods allow for two types of model orientations: input or output (Avkiran and Rowlands 2008). Since our objective is to estimate the relative efficiency of each DMU at generating selected outputs (i.e., reduction of agricultural pressures), we believe the output specification to be the most appropriate. The specific approach used in this study also accounts for the role of undesirable outputs, such as agricultural pressures (quantitative and qualitative) on water resources, in order to assess efficiency among a group of DMUs (e.g., TRBs). Recent DEA applications have shown the relevance of including undesirable outputs (e.g., pollution) as a more realistic specification of the optimization model (Sueyoshi and Goto 2011). Furthermore, recent DEA developments have revealed the importance of using different output specifications (i.e., natural and managerial) in order to obtain additional strategic information for the assessment of DMUs (Expósito and Velasco 2018; Sueyoshi and Goto 2011). With the aim to assess the managerial capacity of RBs at minimizing agricultural pressures, this study uses a managerial output specification.

The managerial specification implies a high methodological sophistication in the sense that a DMU may decrease the directional vector of undesirable outputs by increasing inputs. This type of specification is referred to as "managerial disposability" and is usually related to the implementation of innovative management initiatives (e.g., effective multilateral arrangements to reduce agricultural pressures on water resources at TRB scale). In our specific case, it would reflect the TRB capacity to minimize agricultural pressures on water resources despite a potential increase of agricultural water withdrawals. The optimization model produces an autonomous indicator of relative efficiency referred to herein as "managerial efficiency" scheme. Subsequently, a simplified mathematical description of the applied DEA model is offered. Nevertheless, a detailed description of the DEA model applied can be found in Expósito and Velasco (2018, 2020).

In our DEA model, each j-th DMU $j = 1,\ldots n$, uses inputs $X_j = \left(x_{1j},\ldots,x_{mj}\right)^{\mathrm{T}}$ and generates desirable outputs, represented by $G_j = \left(g_{1j},\ldots,g_{sj}\right)^{\mathrm{T}}$, and undesirable outputs, represented by $B_j = \left(b_{1j},\ldots,b_{hj}\right)^{\mathrm{T}}$. Furthermore, $\mathrm{d}_i^x, i = 1,\ldots m$, $\mathrm{d}_r^g, r = 1,\ldots,s$, and $\mathrm{d}_f^b, f = 1,\ldots,h$ represent slack variables related to inputs, and desirable and undesirable outputs, respectively. $\lambda = \left(\lambda_1,\ldots,\lambda_n\right)^{\mathrm{T}}$ are unknown structural or intensity variables, which are used for connecting the input and output vectors via a convex combination. R is the range resolute throughout the upper and lower bounds of inputs, desirable outputs, and undesirable outputs and is expressed by following expressions:

$$R_i^x = (m+s+h)^{-1}\left(\max\left\{x_{ij} \mid j = 1,\ldots,n\right\} - \min\left\{x_{ij} \mid j = 1,\ldots,n\right\}\right)^{-1}$$

$$R_r^g = (m+s+h)^{-1}\left(\max\left\{g_{rj} \mid j = 1,\ldots,n\right\} - \min\left\{g_{rj} \mid j = 1,\ldots,n\right\}\right)^{-1}$$

$$R_f^b = (m+s+h)^{-1}\left(\max\left\{b_{fj} \mid j = 1,\ldots,n\right\} - \min\left\{b_{fj} \mid j = 1,\ldots,n\right\}\right)^{-1}$$

The managerial efficiency of the k_{th} DMU is evaluated by the following radial model:

$$\text{Max } \xi + \varepsilon\left[\sum_{i=1}^{m}R_i^x d_i^x + \sum_{r=1}^{s}R_r^g d_r^g + \sum_{f=1}^{h}R_f^b d_f^b\right]$$

$$s.t.\sum_{j=1}^{n}x_{ij}\,\lambda_j + (-1)^o\,d_i^x = x_{ik} \quad (i = 1,\ldots,m),$$

$$\sum_{j=1}^{n} g_{rj} \lambda_j - d_r^g - \xi g_{rk} = g_{rk} \quad (r = 1,\dots,s),$$

$$\sum_{j=1}^{n} b_{fj} \lambda_j + d_f^b + \xi b_{fk} = b_{fk} \quad (f = 1,\dots,h),$$

$$\sum_{j=1}^{n} \lambda_j = 1,$$

$$\lambda_j \geq 0 (j = 1,\dots,n), d_i^x \geq 0 (i = 1,\dots,m),$$

$$d_r^g \geq 0 (r = 1,\dots,s), d_f^b \geq 0 (f = 1,\dots,h) \text{ and,}$$

$$\xi : \text{unrestricted} \tag{9.1}$$

Its solution provides the necessary efficiency scores, measured by:

$$\theta^* = 1 - \left[\xi^* + \varepsilon \left(\sum_{i=1}^{m} R_i^x \ d_i^{x*} + \sum_{r=1}^{s} R_r^g \ d_r^{g*} + \sum_{f=1}^{h} R_f^b \ d_f^{b*} \right) \right], \tag{9.2}$$

being $o = 1$.

Once the efficiency mapping of DMUs is estimated, this study tests the impact of certain determinants on the estimated efficiency scores for our sample of European TRBs. The most commonly included determinants (or explanatory factors) in the literature regarding the comparative analysis of the environmental performance of TRBs (see, for example, Bernauer and Kuhn (2010) and Knieper and Pahl-Wostl (2016), among others) are those related with population (e.g., population density), economic development (e.g., per capita income or GDP), overall human pressure on water resources (e.g., water withdrawals on renewable water resources on a RB scale), and geo-location variables. In order to test whether these determinants carry a suffi-ciently large influence to explain efficiency scores in our case study, we have applied both an ordinary least squares (OLS) model, which controls for heteroscedasticity with robust standard errors, and a Tobit model, which controls for upper limit values (i.e., 1 = full efficiency). The use of two alternative regression models contribute to the robustness of the obtained results.

9.3.2 DATA

In 2012, the GEF approved the Transboundary Water Assessment Programme (TWAP) of the United Nations Environment Programme (UNEP) following an earlier programme (Development of the Methodology and Arrangements for the GEF Transboundary Waters Assessment Programme in 2011). TWAP offers a unique homogeneous dataset for TRBs that enables international comparison and assessment studies (UNEP 2016). TWAP constitutes a global assessment platform of 286 TRBs distributed worldwide and offers a wide variety of indicators regarding water quantity, water quality, ecosystems, governance, and socioeconomic fields. This study uses the latest available data (year 2010) of the TWAP dataset (TWAP, 2019). In order to apply our DEA assessment to the 20 European TRBs selected, the following indicators have been employed:

1. Agricultural withdrawals (input indicator): This indicator has been selected as the necessary input for agricultural activities (i.e., irrigation and livestock production). This input is withdrawn from the RB system (both surface and groundwater sources) for production objectives and given back (partially) in the form of returns which usually contain pollutants, such as nitrogen and phosphorus compounds. Values are measured in $km^3/year$ (Table 9.3).
2. Nutrient pollution (qualitative output indicator): This indicator considers river pollution from nitrogen and phosphorus compounds, which are mainly caused by agricultural activities (Bernauer and Kuhn 2010; Dimitriou et al. 2012; EEA 2012b). Urban wastewater and atmospheric deposition of nitrogen can also generate nutrient pollution, but to a much lesser extent (Boyle 2014). The construction of the indicator is based on the methodology developed by Mayorga et al. (2010) and Seitzinger et al. (2010). In our specific DEA optimization model, this indicator needs to be minimized. Values range from 0 to 1 (from less to more highly polluted) (Table 9.3).
3. Agricultural abstraction stress (quantitative output indicator): This indicator identifies agricultural water stress, and is constructed based on the consumption-to-availability ratio (mean annual water consumption divided by the sum of mean annual runoff on a river basin scale). A reduction of the agricultural stress indicator implies a decrease in the pressure of agriculture on water resources of the RB (e.g., through more efficient practices of water use). Its introduction as an additional output indicator is justified since the presence of agricultural nutrient pollution in RBs is strongly related to quantity factors, as measured by consumption-to-availability ratios (EEA 2012b;

TABLE 9.3
Descriptive Statistics

	Agricultural withdrawals	Nutrient pollution	Abstraction stress
Average	3,688	0.69	0.06
St. Deviation	4,589	0.21	0.07
Maximum	19,397	1	0.30
Minimum	0.83	0.25	0.00

Munia et al. 2016). In our specific case, this output has been introduced in terms of its inverse value, therefore this indicator needs to be maximized in our optimization problem. Values range from 0 to 1 (Table 9.3).

In summary, our optimization problem aims to minimize not only the agricultural nutrient pollution (qualitative output indicator) but also the agricultural abstraction stress (quantitative output indicator). Therefore, the DEA method applied enables both output indicators to be simultaneously optimized. Descriptive statistics of our input and output variables are shown in Table 9.3. It is worth noting that the contrast of output/input ratios of each RB with the mean values shows that any potential problems of sensitivity caused by the existence of outliers and statistical noise are ruled out. DEA methods do not require data to be normalized (e.g., by agricultural activity or withdrawals in the basin) to obtain efficiency scores. Additionally, the condition regarding the minimum number of observations per variable established by Banker et al. (1996) is met (20 DMUs and 3 input–output variables). Hence no misspecification problems are observed.

9.4   RESULTS

As discussed in the previous section, relative efficiency relies on effective management initiatives to minimize undesirable outputs. In our managerial output specification, the imposition of the reduction of undesirable outputs can be achieved despite an input increase (i.e., higher water withdrawals). Under this assumption, the obtained efficiency mapping describes the capacity to reduce agricultural pressures on water resources (despite potential higher withdrawals of water for agriculture) and helps to identify those benchmark RBs from which to learn.

The results of our efficiency assessment for our sample of European RBs are presented in a step-by-step manner. The first step focuses on the efficiency mapping obtained from our model specification (Table 9.4). The results identify four efficient (benchmark) RBs: the Danube, Volga, Kemi, and Vuoksi RBs. In these RBs, an increase in agricultural withdrawals would not necessarily imply greater pressures on water resources from agriculture. This can be explained by the implementation of effective management initiatives that positively impact on the input–output intrinsic relationship. Conversely, inefficient RBs would therefore need to increase their management efforts toward reducing agricultural pressures, since the increase of agricultural withdrawals would be counterproductive. Within the heterogeneous range of inefficient RBs, the Elbe and Rhine RBs register the lowest efficiency scores (0.25), thus showing the urgent need to implement agricultural management initiatives at basin scale. Interestingly, the Kemi and Vuoksi RBs, two efficient DMUs, present a very low proportion of agricultural water withdrawals on total withdrawals on a basin scale, which could be argued as a possible explanatory factor for its high efficiency scores (since lower agricultural water withdrawals would lead to less agricultural nutrient pollution and less abstraction stress). Nevertheless, the Volga and Danube RBs register greater agricultural water withdrawals (Table 9.1) and are also fully efficient DMUs in our managerial specification. Therefore, as widely argued in the existing literature (Zhu 2016), DEA outcomes do not depend on input data (or on its

TABLE 9.4
Mapping of Efficiency Scores

	Efficiency score
Danube	1
Dnieper	0.64
Don	0.34
Douro	0.45
Ebro	0.50
Elbe	0.25
Guadiana	0.55
Kemi	1
Klarälven	0.57
Maritsa	0.35
Neman	0.42
Oder	0.33
Po	0.52
Rhine	0.25
Rhone	0.33
Seine	0.52
Tagus	0.25
Vistula	0.33
Volga	1
Vuoksi	1
Average	*0.54*
St. Dev.	*0.26*
EU²	*0.38*
Non-EU³	*0.82*

Note: [1]Weighted by drainage area. [2]EU group with more than 50 percent of the TRB drainage area within EU countries. [3]Conversely, more than 50 percent of the TRB area within non-EU countries.

heterogeneity among DMUs), but instead on the intrinsic input–output relationships optimized for the selected group of DMUs. Furthermore, the less efficient DMUs, the Elbe and Rhine RBs, show a low proportion of agricultural withdrawals of the total, thus showing that a low input value does not necessary lead to a high efficiency score. Additionally, it is worth noting that average values show a relatively low level of efficiency, 0.54 (Table 9.4), and there is high heterogeneity in our sample (as measured by the estimated standard deviations).

In a final step, the analysis focuses on the determinants of our estimated efficiency scores. To this end, an OLS model is estimated, which controls for heteroscedasticity with robust standard errors (Table 9.5), together with a Tobit model, which controls for upper limit values (i.e., 1 = full efficiency) (Table 9.6). The results show whether certain factors (as described in Section 3.1) determine the efficiency scores achieved by our group of RBs. Both estimation techniques show similar estimated parameters in terms of the signs of the relationship and significance levels, thus

TABLE 9.5
Analysis of Determinants of Efficiency (OLS Robust)

Variable	Coeff.	p-value
GDP per capita[1]	0.108	0.094
Non-EU[2]	0.029	0.816
No. of countries	-0.024**	0.014
Population density[1]	-0.151***	0.002
RB Area[1]	0.147**	0.026
TRBMIs [2]	-0.027	0.121
Withdrawals (over total renewable resources)[1]	0.001	0.988
R^2	0.721	

Note: [1]Variables in logs. [2]Binary variables. *** Denotes significance at 1 percent. ** Denotes significance at 5 p.

TABLE 9.6
Analysis of Efficiency's Determinants (Tobit Model)

Variable	Coeff.	p-value
GDP per capita[1]	0.134	0.141
Non-EU[2]	0.076	0.687
No. of countries	−0.034*	0.084
Population density[1]	−0.176***	0.005
RB Area[1]	0.174***	0.009
TRBMIs [2]	−0.032	0.278
Withdrawals (over total renewable resources)[1]	−0.009	0.835
LR Chi[2]	26.11	

Note: [1]Variables in logs. [2]Binary variables. *** Denotes significance at 1 percent. * Denotes significance at 10 percent.

showing sufficient robustness of our estimates. No multicollinearity problems have been detected.

Estimates show that variables, such GDP per capita (measured in US dollars), non-EU versus EU (binary variable that takes value 1 when more than 50 percent of the RB area is located in non-EU countries), the existence of transboundary river basin management institutions (TRBMIs) (binary variable that takes value 1 if a management transboundary organization or commission exists) and total water withdrawals are revealed to be non-significant and therefore fail to explain efficiency scores. On the other hand, as expected, factors, such as an increasing number of countries, population density (inhabitants per square kilometer) and RB area, exert a negative effect on efficiency scores.

Based on the obtained results, it seems that EU RBs (understood as those mostly located in EU territory) fail to perform better as a result of the significant efforts derived from intensive water and agricultural regulation, compared to RBs mainly located in non-EU countries (Table 9.4). In this respect, and following Bernauer and Kuhn (2010), it is argued that the absence of any effect of EU membership on better management of agricultural pressures may indicate that agriculture and water policies are not sufficiently coordinated, and that the EU Common Agricultural Policy (CAP) might not being providing the results expected in terms of reducing intensive farming based on high use of water and fertilizers. This might also be understood as EU RBs had more intensive agriculture than non-EU RBs. Another interesting result shows that the existence of TRBMIs fails to exert a positive effect on the estimated efficiency scores, and therefore the mere creation of transboundary multilateral institutions does not necessarily lead to a better performance in terms of reducing agricultural pressures (both quantitative and qualitative) on water resources at RB scale. Again, the lack of effective cooperation and of strategic policy coordination could lie behind these results. Finally, the nonexistence of an income effect (as measured by GDP per capita) on efficiency mappings would show that agricultural pressures are less sensitive to capital-intensive factors (since richer countries can more easily implement measures such as treatment plants, irrigation modernization programs, and agricultural compensation schemes) and therefore require greater political and cooperation efforts (Bernauer and Kuhn 2010; Expósito et al. 2019). These findings are also in line with those of Dinar (2008) and of Knieper and Pahl-Wostl (2016). Additionally, a high number of riparian countries seem to interfere negatively on the RB performance at reducing undesirable pressures of agricultural activity on water resources, thereby achieving lower efficiency scores. In this respect, Dinar (2008) argues that the larger the number of states involved in transboundary river cooperation becomes, the more difficult it is to achieve effective cooperation for agricultural pollution abatement. Similarly, higher population densities constitute increasing pressures on the RB that would explain the observed negative effect on efficiency scores. Again, it is worth noting the case of the Danube RB, since it registers a high number of riparian countries and population density, as well as relatively high agricultural water withdrawals. Despite these determinants, the Danube RB is flagged as a benchmark unit due to its exemplary management initiatives. Finally, the RB area seems to play a positive role in explaining efficiency scores, thereby indicating that RBs of a more extensive nature would achieve better performance results. This result could be related to the consumption-to-availability ratio on a RB scale, since greater (in area) RBs would register greater renewable resources and would therefore register lower pressure of water withdrawals.

The findings offer highly relevant information to policy and decision makers, since they clearly suggest that current managerial initiatives are insufficient to achieve a sustainable agriculture, as well as to reduce the negative agricultural pressures on water resources. Therefore, a more effective transboundary cooperation is required and innovative initiatives focused on reducing agricultural pollution and efficient agricultural water use (thus reducing abstraction pressure) need to be implemented. Managerial initiatives implemented by the benchmark basins identified in this study, as well as the identification of the factors affecting the managerial efficiency to reduce agricultural pressures, might offer valuable information for decision makers.

9.5 CONCLUSION

Transboundary rivers are under constant strain to meet the demands from agriculture and other uses. In this context, the joint efforts of all stakeholders implied in RB management, as required by the WFD and the Aarhus Convention, remain necessary for the development and support of the implementation of effective management measures to address agricultural pressures on water resources and achieve a more sustainable agriculture. Despite existing limitations, the Danube RB and its ICPDR constitute a good example to learn from, since our DEA analysis shows that it is an efficient managerial unit. Additionally, our results suggest that, at least in the case of the selected European RBs, the capacity to reduce agricultural negative pressures on water resources is more closely related to RB characteristics (e.g., area, population density, number of riparian countries) than to policy or institutional factors (e.g., EU membership, existence of TRBMIs). Findings offer highly relevant information to policy and decision makers, since they clearly suggest that current managerial initiatives are insufficient to achieve a sustainable agriculture and reduce the negative agricultural pressures on water resources. A more effective transboundary cooperation is required and initiatives focused on reducing agricultural pollution and efficient agricultural water use (thus reducing abstraction pressure) need to be implemented. In this sense, the Danube river basin offers some good learning examples of effective multi-country and multi-agent cooperation to address agricultural pressures, such as the Environmental Programme for the Danube River Basin (EPDRB) and the "Friends of the Danube" programme.

Finally, it is worth noting that the analysis carried out in this paper is of a static nature. Future research will focus on carrying out a dynamic analysis, which could offer additional significant information for the adaptation of RB management initiatives to a changing environment. Nevertheless, the TWAP data base has not been up-dated, being the latest data referred to year 2010, what constitutes a limitation to carry out a dynamic analysis. As soon as more data becomes available, we aim to evaluate dynamic changes in efficiency mappings and the factors that may explain these changes.

REFERENCES

Avkiran N, Rowlands T (2008) How to better identify the true managerial performance: state of the art using DEA. *Omega-Int J Manage S* 36(2):317–324.

Bakker MHN, Duncan JA (2017) Future bottlenecks in international river basins: where transboundary institutions, population growth and hydrological variability intersect. *Water Int* 42(4):400–424.

Banker RD, Chang H, Cooper W (1996) Simulation studies of efficiency, returns to scale and misspecification with nonlinear functions in DEA. *Ann Op Res* 66(4):231–253.

Banker RD, Charnes A, Cooper W (1984) Some models for estimating technical and scale inefficiencies in data envelopment analysis. *Manag Sci* 30(9):1078–1092.

Berbel J, Expósito A (2018) Economic challenges for the EU Water Framework Directive reform and implementation. *Eur Plan Stud* 26(1):20–34.

Bernauer T, Kuhn PM (2010) Is there an environmental version of the Kantian peace? Insights form water pollution in Europe. *Eur J Int Relat* 16:77–102.

Boyle S (2014) The case for regulation of agricultural pollution. *Environ Let Rev* 16:4–20.

Charnes A, Cooper W, Rhodes E (1978) Measuring the efficiency of decision making units. *Eur J Oper Res* 2(6):429–444.

Charnes A, Cooper W, Rhodes E (1981) Evaluating programme and managerial efficiency: an application of data envelopment analysis to Programme Follow Through. *Manag Sci* 27(6):668–697.

Dimitriou E, Mentzafou A, Zogaris S, Tzortziou M, Gritzalis K, Karaouzas I, Nikolaidis C (2012) Assessing the environmental status and identifying the dominant pressures of a trans-boundary river catchment, to facilitate efficient management and mitigation practices. *Environ Earth Sci* 66(7):1839–1852.

Dinar A (2008) *International Water Treaties. Negotiation and Cooperation along Transboundary Rivers*. New York: Routledge.

Dinar A, Dinar S, McCaffey S, McKinney D (2013) *Bridges over Water. Understanding Transboundary Water Conflict, Negotiation and Cooperation*. New Jersey: World Scientific.

EC (2012) A blueprint to safeguard Europe's water resources. Communication from the Commission to the European Parliament, the European Economic and Social Committee and the Committee of the Regions (COM (2012) 673). European Commission, Brussels.

EEA (2012a) European waters. Assessment of status and pressures. Report no 8/2012. Environment European Agency, Copenhagen.

EEA (2012b) European waters. Current status and future challenges. A synthesis. Report no 9/2012. Environment European Agency, Copenhagen.

EEA (2015) The European Environment. State and Outlook 2015. Synthesis Report. Environment European Agency, Copenhagen.

EU (2017) Agri-environmental schemes: how to enhance the agriculture-environment relationship. Science for Environment Policy. Thematic Issue 57. Issue produced for the European Commission DG Environment by the Science Communication Unit, UWE, Bristol.

Expósito, A, Pablo-Romero, MP, Sánchez-Braza, A (2019). Testing EKC for urban water use: Empirical evidence at river basin scale from the Guadalquivir River, Spain. *J Water Resour Plan Manag* 145: 4.

Expósito A, Velasco F (2018) Municipal solid-waste recycling market and the European 2020 Horizon Strategy: a regional efficiency analysis in Spain. *J Clean Prod* 172:938–948.

Expósito A, Velasco F (2020) Exploring environmental efficiency of the European agricultural sector in the use of mineral fertilizers. *J Clean Prod* 253:119971.

FAO (2017) *FAOSTAT Food and Agricultural Data*. Food and Agriculture Organization of the United Nations. Rome, Italy.

Farrell, M. J. (1957). The measurement of productive efficiency of production. *J R Stat Soc A* 120(III), 253–281.

Gerlak AK (2004) The global environment facility and transboundary water resource management: new institutional arrangements in the Danube river and Black Sea region. *J Environ Dev* 23(4):400–424.

Giordano M, Drieschova A, Duncan JA, Sayama Y, De Stefano L, Wolf AT (2014) A review of the evolution and state of transboundary freshwater treaties. *Int Environ Agreem-P* 14(3):245–264.

Jager NW (2016) Transboundary cooperation in European water governance. A set-theoretic analysis of international river basins. *Environ Policy Gov* 26:278–291.

Kallioras A, Pliakas F, Diamantis I (2006) The legislative framework and policy for the water resources management of transboundary rivers in Europe: the case of Nestos/Mesta River, between Greece and Bulgaria. *Environ Sci Policy* 9:291–301.

Kauffman GJ (2015) Governance, policy, and economics of intergovernmental river basin management. *Water Resour Manag* 29:5689–5712.

Knieper C, Pahl-Wostl C (2016) A comparative analysis of water governance, water management, and environmental performance in river basins. *Water Resour Manag* 30:2161–2177.

Libert B (2015) The UNECE Water Convention and the development of transboundary cooperation in the Chu-Talas, Kura, Drin and Dniester river basins. *Water Int* 40(1):168–182.

Mayorga E, Seitzinger SP, Harrison JA, Dumont E, Beusen AHW, Bouwman AF, Fekete BM, Kroeze C, Van Drecht G (2010) Global nutrient export from WaterSheds 2 (NEWS 2): Model development and implementation. *Environ Modell Soft* 25:837–853.

Mohaupt V, Crosnier G, Todd R, Petersen P, Dworak T (2007) WFD and agriculture activity of the EU: first linkages between the CAP and the WFD at EU level. *Water Sci Technol* 56(1):163–170.

Munia H, Guillaume JHA, Mirumachi N, Porkka M, Wada Y, Kummu M (2016) Water stress in global transboundary river basins: significance of upstream water use on downstream stress. *Environ Res Lett* 11:014002.

Ostrom E (2005) *Understanding Institutional Diversity.* Princeton, PA: Princeton University Press.

Özerol G, Bressers H, Coenen F (2012) Irrigated agriculture and environmental sustainability: an alignment perspective. *Environ Sci Policy* 23:57–67.

Petersen-Perlman JD, Veilleux JC, Wolf AT (2017) International water conflict and cooperation: challenges and opportunities. *Water Int* 42(2):105–120.

Pulwarty RS, Maria R (2015) Adaptation challenges in complex rivers around the world: the Guadiana and the Colorado basins. *Water Resour Manag* 29:273–293.

Romano G, Guerrini A, Marques RC (2017) European water utility management: Promoting efficiency, innovation and knowledge in the water industry. *Water Resour Manag* 31:2349–2353.

Seitzinger SP, Mayorga E, Bouwman AF, Kroeze C, Beusen AHW, Billen G, et al. (2010) Global river nutrient export: A scenario analysis of past and future trends. Global Biogeochem Cy 24:GB0A08.

Sueyoshi T, Goto M (2011) Measurement of returns to scale and damages to scale for DEA-based operational and environmental assessment: how to manage desirable (good) and undesirable (bad) outputs? *Eur J Oper Res* 211(1):76–89.

Sueyoshi T, Sekitani K (2009) An occurrence of multiple projections in DEA-based measurement of technical efficiency: theoretical comparison among DEA models from desirable properties. *Eur J Oper Res* 196(2):764–794.

TWAP (2019) Transboundary Water Assessment Programme. River Basins Component. UNEP and GEF. Available at http://twap-rivers.org

UNECE (1992) Convention on the protection and use of transboundary watercourses and international lakes. Helsinki, March 17, 1992, UN Economic Commission for Europe, United Nations.

UNECE (2011) Second assessment of transboundary rivers, lakes and groundwaters. EDE/MP.WAT/33, UN Economic Commission for Europe, United Nations.

UNEP (2016) Transboundary water systems – status and trends: crosscutting Analysis. United Nations Environment Programme (UNEP), Nairobi.

Van Rijswick M, Gilissen HK, Van Kempen J (2010) The need for international and regional transboundary cooperation in European river basin management as a result of new approaches in EC water law. *ERA Forum* 11: 129–157.

Wiering M, Verwijmeren J, Lulofs K, Feld C (2010) Experiences in regional cross border cooperation in river management. Comparing three cases at the Dutch–German border. *Water Resour Manag* 24(11):2647–2672.

Xiang Z, Chen X, Lian Y (2016) Quantifying the vulnerability of surface water environment in humid areas based on DEA method. *Water Resour Manag* 30:5101–5112.

Zawahri NA, Mitchell SM (2011) Fragmented governance of international rivers: Negotiating bilateral versus multilateral treaties. *Int Stud Quart* 55(3):835–858.

Zhu J (2016) *Data Envelopment Analysis: A Handbook of Empirical Studies and Applications.* New York: Springer.

10 The Viability of Nuclear Power as an Alternative to Renewables for Clean Energy for Climate Change Mitigation

Beth-Anne Schuelke-Leech and Timothy C. Leech

10.1 INTRODUCTION

The need for sustainable, non-carbon power sources has never been greater. Climate change is an accelerating global crisis. It cannot be effectively addressed without cleaner energy sources that displace the use of fossil fuels. Energy from renewable sources, such as solar, wind, geothermal, hydro, and biomass, provides communities and electric utilities with "green" energy, that is, zero-emissions energy coming from sustainable sources. Renewable energy capacity has been growing significantly in recent years, with an increase of 10 percent globally in 2020 alone, or approximately

DOI: 10.1201/9781003240129-10

107 GW, bringing total global capacity to 1,398 GW [1]. In the United States, renewable energy will account for almost 70 percent of new generating capacity in 2021, with installed capacity of 284.6 GW providing 23.4 percent of the electricity [2].

At the same time, renewables have several problems with their widespread deployment [3]. They are dependent on the right conditions for harvesting energy (e.g., the wind must be blowing, or the sun must be shining), which means that renewables are variable (i.e., intermittent) energy sources. Thus, they require either some storage capacity or else a baseload power source to cover energy demands when they are not available. Energy storage via batteries have environmental costs, as do renewable power technologies (e.g., wind turbines, photovoltaic panels, geothermal piping, etc.) themselves. Batteries require significant amounts of metals and non-metals, which often have to be mined in remote or constrained areas [4]. The lifecycle costs of solar power are actually slightly higher than the lifecycle costs of nuclear power, while those of wind power are slightly less [5].

Renewable energy requires significant other resources, as well. For instance, wind and solar farms take a significant amount of land [3]. A study led by Pimentel estimated that renewables could provide almost 50 percent of US energy needs, but this would require approximately 17 percent of the nation's land area [6]. A more recent study showed that for solar energy to provide one-third of US energy needs, it would take approximately 10,000 square kilometers of real estate [7], approximately two times the size of Delaware or 83 percent the size of Connecticut [8]. Providing the same amount of energy from wind power would require the allocation of approximately 66,000 square kilometers of land [7], slightly more than the area of New Hampshire, Vermont, and Massachusetts combined [8]. With both of these sources providing this energy, they would still only provide two-thirds of the energy needs of the United States. There would remain another one-third of energy to produce. This real estate is not readily available. The best land for renewable power is not necessarily in remote "uninhabited" locations. Solar power, for instance, is best located in southern latitudes with plenty of direct sunlight. However, these parts of the United States are also populated. Displacing people for renewable energy installations is likely to be unacceptable. In addition to disrupting human habitat and agricultural areas, significant renewable energy installations are likely to disrupt wildlife habitat, potentially leading to greater human vulnerability through disease.[1] As mentioned above, mining metals and non-metals have additional impacts on land [4].

The final concern with renewable energy is the cost [3]. Evidence suggests that renewables are increasing the price of electricity. For example, electricity prices in California rose by 28 percent between 2011 and 2018, seven times more than the national average of 5 percent. Electricity prices in Germany have risen by 50 percent since 2006, largely as a result of the adoption of renewable energy [10]. In some states in the United States, such as Ohio, legislatures are looking to nuclear power as an alternative to renewables [10]. In many countries, the deployment of renewables has been aided by subsidies that have lowered costs to utilities and customers. Though the costs are decreasing, there are concerns that the deployment of renewable energy may stall as governments announce ends to subsidies. For instance, China has announced that renewables will have to openly compete with fossil fuels in the energy market

[11]. In the United States, subsidies are due to expire at the end of 2021, though renewal is possible [10].

Though not strictly a renewable energy, nuclear power is akin to renewables in its ability to provide energy without greenhouse gas emissions. It can provide emission-free, clean baseload energy [12]. However, nuclear power has many issues that make it questionable as a sustainable energy source. Innovation has the potential to solve many of the problems with nuclear power. However, it is unclear whether these innovations can be developed and implemented within the next 20 years, when significant progress toward a zero-carbon energy grid is essential in order to escape the direct impacts of rapid climate change.

This chapter examines nuclear technology innovation and industry regulation in order to understand the difficulties of relying on nuclear power as a meaningful alternative to renewable power deployment. It considers the current status of nuclear energy innovations and the extent to which the progress and obstacles for nuclear power supports including it as a viable alternative, or supplement, to renewable energy. We extend this introduction by providing an overview of current challenges within the nuclear energy sector and briefly surveying nuclear reactor designs and innovations. Then, we discuss the process of nuclear design and development within the United States, with an emphasis on the role of the Nuclear Regulatory Commission (NRC). The results of an empirical study on discussions by the NRC on Gen III and Gen IV reactors are presented. Finally, we discuss the implications for the urgent need to address climate change.

10.2 NUCLEAR POWER

Nuclear power has been an important source of energy for the past 50 years. Nuclear power currently provides a significant portion of the electricity in many countries: 71 percent in France, 48 percent in Belgium, 34 percent in Sweden, 49 percent in Hungary, 26 percent in South Korea, 20 percent in the United States, 20 percent in Russia, 16 percent in the United Kingdom, and 15 percent in Canada [13]. Nuclear reactors in the United States currently provide 50 percent of the carbon-free energy [14].

Nuclear reactors are generally classified into four types. The first is Generation I (Gen I), which were the prototypes and reactors originally developed in the 1950s and 1960s. Gen II reactors are the reactors currently deployed and operating around the world. They were designed and built in the 1960s through the 1990s [15]. Gen III reactors are enhanced Gen II reactors, with greater thermal efficiency, modularized construction, and improved safety systems [56]. They were developed in the 1990s to have standardized, simpler designs, passive safety features, greater fuel efficiency to reduce refueling and spent fuel, and a longer design life [16]. Gen III+ reactors are Gen III reactors with enhanced safety and smaller production capacity able to be sequenced and combined (called small modular reactors). These are the most advanced reactors currently being deployed for large-scale production [17].

Gen IV reactors are the next iteration of advanced nuclear reactors [18].[2] Nuclear power capacity has been increasing, with approximately 50 reactors currently under construction [19]. Most of this expansion is occurring in Asia and Russia. Many of the existing reactors in the United States and those in other developed countries such as the UK, Canada, and France, are reaching the limits of their operating lives and will need to be replaced with some form of energy production. Most of the retiring nuclear power plants in the United States are typically replaced with energy produced by either coal or natural gas [20]. Significant capacity is being created by extending the life of nuclear plants and increasing their capacity through changes to the thermal cycle, called uprates. In the United States, the NRC, the federal regulator, has approved 165 uprates since 1977, yielding in an increased capacity of 7,500 MWe [19].

New nuclear reactor designs are proposed to be much safer, without the long-term legacy problems of radioactive waste [21–23]. However, most of these reactor designs are still in early phases of development. While the People's Republic of China and Russia are developing and constructing new nuclear reactors, other jurisdictions such as the United States, Canada, and the European Union struggle with their nuclear policies and developing new nuclear technologies.

Nuclear power remains a contentious source of energy. The problems of nuclear power are well-documented. Light water reactor technologies—by far the most common reactors operating today—have inherent problems that have proven difficult to solve, including safety, waste management, and cost overruns. They also face significant public backlash due to these problems. The nuclear industry itself is burdened with cost overruns, safety concerns, and waste management problems [24].

Meltdowns and accidents are a major public concern about the nuclear power industry. In 1990, the International Atomic Energy Agency (IAEA) and the Nuclear Energy Agency (NEA) of the Organization for Economic Cooperation and Development (OECD) jointly developed the International Nuclear and Radiological Event Scale (INES), which was designed to consistently identify nuclear and radiological events [25]. The INES scale goes from Level 1 (for anomalies) to Level 7 (major accidents). The global nuclear industry has experienced two Level 7 events since 1970: The meltdown and explosion at Chernobyl in the Ukraine in 1986 and the tsunami-induced meltdown at the Fukushima Daiichi power plant in March 2011 [26]. The accident that occurred in 1979 at the Three Mile Island power plant, resulting in the partial meltdown of one of its reactors was rated at a Level 5 [27]. This incident marked a major turning point in public perception of nuclear safety in the United States [28]. These accidents demonstrate that meltdowns, and the corresponding social, environmental, political, and economic impacts, are both possible and substantial [29]. However, nuclear power generation also has a strong safety record over its history of operations, with the lowest deaths from energy-related accidents per unit of energy produced of any source of energy, including wind power and solar energy [30].

Nuclear power plants are far more complex than conventional fossil fuel plants. Where a typical fossil fuel plant has roughly 4,000 valves and 5,000 pipe supports, a typical nuclear plant might have ten times the number of valves, and four or five times the number of pipe supports [31]. This complexity is one of the things that makes nuclear power plants riskier and in greater need of inherent system-wide safety [32]. Complex systems are more likely to be brittle and susceptible to failures and problems [33].

The capital costs of reactor development, licensing, and construction make it unattractive economically and are major impediments to developing new nuclear power plants [18,34]. Advanced nuclear power has approximately four times the estimated capital costs for coal, based on the current average size of the power plant (see Table 10.1).

Even for the same sized plant, construction costs for advanced nuclear are forecast to be more than every other type of energy. Though variable operating costs for advanced nuclear is relatively low, it is still higher than wind, solar photovoltaics, and geothermal power. Advanced nuclear also has relatively high fixed operating and maintenance costs. Its track record of consistent and volatile cost overruns and project delays make nuclear power a problematic investment. Thus, advanced nuclear has a significant market disadvantage.

Construction cost overruns and delays are major obstacles. Construction of nuclear power plants frequently takes many years. The average construction time for the 37 reactors started globally since 2004 is ten years, twice as long as is typically forecast at the start of the projects [37]. Of the 53 reactors currently under construction worldwide, 37 of them are behind schedule [38]. Cost overruns, ongoing delays, and numerous increasing budget forecasts create barriers for the adoption of nuclear reactors [39].

The events of the tsunami that ultimately led to nuclear meltdowns at the Fukushima Daiichi power plant in 2011 renewed fears of nuclear accidents and decreased public support for nuclear power [40]. Gen III reactors were supposed to be simpler and less expensive, but the financial challenges for both Westinghouse and Areva show that Gen III reactors are running into the same escalating costs and schedule delays [41]. This makes it hard for policymakers, utility companies, and the general public to trust nuclear construction forecasts, budgets, and technologies. The nuclear industry has consistently been overly optimistic about the forecasts for construction times and budgets. Thus, it is entirely reasonable that the promises of advanced nuclear reactors are viewed with considerable skepticism.

Solutions for the problems with nuclear power are certainly needed if nuclear power is to contribute to a sustainable energy future. Gen III and III+ reactors aimed to increase safety by incorporating some design features developed after the deployment of Gen II reactors. One important enhancement is passive cooling. Gen II reactors required the active cooling of reactors from electricity-driven pumps. If power is lost, the pumps can fail and the coolant system can be lost.[3] Passive cooling requires no action from an operator for the plant to shut down in the event of an emergency[4] [43]. Thus, it is viewed as much safer for operations. Another important feature was standardization and modularization. Small modular reactors (SMR) are designed to be relatively small (typically 50–100 MWe), closed systems where multiple SMRs are used together (modularity), rather than relying on a single mega-unit. The advantage of SMRs is that the utilities can lower their capital costs and increase design certainty and safety [44]. In September 2020, a SMR from NuScale was approved by the US NRC, the first SMR approved by the NRC after a lengthy design and development process [45].

No country has yet solved the problem of spent fuel (i.e., nuclear waste) entirely.[5] Currently, spent fuel is stored onsite at each nuclear power plant since there is no

TABLE 10.1
Costs of Electricity Generation by Source in the United States (in 2016$)[a]

Source	Average size (MW)	Construction cost (per kW)	Baseline construction costs (in $000)	Variable operating costs (per MW hour)	Fixed operating and maintenance costs (per kW/ year)	Levelized[b] electricity cost per MW hour	Baseline construction costs for 2,000 MW plant (in $000)
Natural gas	702	$969	$680,238	3.48	10.93	$140.00	$1,938,000
Wind	100	$1,686	$168,600	0	46.71	$57.50	$3,372,000
Solar photovoltaic	100	$2,277	$227,700	0	21.66	$99.10	$4,554,000
Coal with 30% carbon sequestration	650	$5,030	$3,269,500	7.06	69.56	$52.20	$10,060,000
Advanced nuclear	2,234	$5,880	$13,135,920	2.29	99.65	$66.80	$11,760,000
Hydropower	500	$2,442	$1,221,000	2.66	14.93	$66.20	$4,884,000
Geothermal	50	$2,715	$135,750	0	117.95	$43.30	$5,430,000
Biomass	50	$3,790	$189,500	5.49	110.34	$102.40	$7,580,000

[a] From [35]. U.S. Energy Information Administration. (2017, December 30, 2017). Cost and Performance Characteristics of New Generating Technologies, *Annual Energy Outlook 2017*. Available: www.eia.gov/outlooks/aeo/assumptions/pdf/table_8.2.pdf and [36] U.S. Energy Information Administration. (2016, December 30, 2017). Levelized Cost and Levelized Avoided Cost of New Generation Resources, *Annual Energy Outlook 2017*. Available: www.eia.gov/outlooks/aeo/pdf/electricity_generation.pdf

[b] Levelized Cost of Electricity is the total costs over the lifetime of the power plant divided by the total electrical energy produced over that lifetime

domestic repository in the United States [48]. Every country supporting the development of advanced nuclear reactors has a stated goal of a closed fuel cycle, which would eliminate the problem of spent fuel [see 13,30,49,50–52]. However, this goal is still theoretical.

Progress on some of the other problems with nuclear power, such as the spent fuel problem, will not come until more advanced reactors are designed and deployed. However, Gen IV reactor designs are incomplete and commercially unproven. New reactors are costly to design, particularly when it is uncertain how much of a market will exist for these new power sources. The returns for any investments are uncertain, which is one of the reasons that public sector investments in these technologies are critical for their development. As discussed in the next section, nuclear power innovations are proposed and under development in many countries around the world. We also discuss the viability of these designs.

10.3 NUCLEAR POWER INNOVATION

Innovative nuclear reactor designs are touted as solutions to many of the problems that exist with the nuclear energy industry. The next generation nuclear reactors are forecast to be inherently safer, with passive safety systems [53].

One of the underlying characteristics of innovation is a willingness to experiment and take on risks in unproven design. The cost of development for advanced reactors is significant and companies cannot afford to undertake these development activities when the return on them is uncertain.

Nuclear reactor designs are extraordinarily expensive and cannot be easily modified. The design becomes "locked-in" 12–15 years before actual operation,[6] which can then last from 40–80 years. Thus, a lack of regulator engagement increases the uncertainty during research and development (R&D). Without regulatory approval, all design and R&D work could come to a naught. There is no way to recoup investments until the design has been licensed and deployed. Long lead times mean that significant capital resources and managerial contingencies need to be factored into the design process. The regulatory and economic constraints mean that companies have little chance to test and improve their design on an iterative basis because of the costs associated with changes. Thus, innovation is too expensive. The technical and regulatory challenges substantially increase the risks and costs for nuclear reactor designs and power plant construction.

Four new nuclear power plants in the United States were announced in the 2000s [57]—two in South Carolina and two in Georgia—which were to use the AP1000 from Westinghouse.[7] However, in 2019, the South Carolina Electric and Gas Company announced that they were halting construction on the two new nuclear plants in the state. The two Gen III+ reactors had faced significant cost overruns and increasing competition from cheap natural gas, making the estimated additional $15 billion required for completion uneconomical [58].

Companies based in developed nations, such as Westinghouse, Areva, Hitachi, and Mitsubishi Heavy Industries, have led in nuclear technologies in the past. However, both Westinghouse and Areva have struggled with recent construction projects. Areva began building a reactor in Finland in 2005 with a forecast cost of €3.2 billion and a target completion of 2009. Cost overruns and delays initially pushed back the target completion to 2018 at a cost of €8.5 billion [59]. It was then delayed again, currently targeted for completion in February 2022 [60]. Westinghouse had been responsible for the construction of the new nuclear power plants in the United States and had experienced problems, resulting in a $6.1 billion loss for the company. The company filed for bankruptcy in early 2017 [61]. Westinghouse sold off its nuclear division to Toshiba, which subsequently sold it to a private equity firm, Brookfield Business Partners, in 2018 [62].

In the United States, nuclear power production has fundamentally used the same technology since the 1950s. Despite the fact that other reactor designs have been proposed (and some have even been tested), light water reactors (either boiling or pressurized) remain the dominant designs. They are the only designs in commercial operations in the United States [63]. Advanced reactors (Gen IV reactors) are not forecast to be deployed until 2030–2050 [18]. Though there are proponents of these technologies, the cost of developing and regulating nuclear reactors makes it virtually impossible for any private entity or company to undertake this on their own [64]. Thus, the development of new nuclear reactor designs has generally been done by the public sector, either through the military as was done in the United States, or through partnerships with private sector organizations [65]. Developing nuclear reactors are long-term projects that have hitherto been undertaken as strategic public investments.

Gen IV reactors are still being designed and developed. None have been commercially deployed. As yet, there is no consensus on the optimal technology for the mass deployment of Gen IV reactors. In 2000, nine countries formed the Generation IV International Forum (GIF) to identify and advance Gen IV reactor technologies [66]. They identified six potential designs: very high-temperature reactor (VHTR); sodium fast reactor (SFR); supercritical water-cooled reactor (SCWR); gas-cooled fast reactor (GFR); lead-cooled fast reactor (LFR); and molten salt reactor (MSR) [67].

A *very high-temperature reactor* (VHTR), or *high-temperature gas-cooled reactor* (HTGR), is a thermal reactor cooled by flowing gas [68]. The high temperature of the coolant (up to 1,000 °C) enables high thermal efficiency of the reactor, which is desirable for high thermal energy applications and industrial co-generation [69]. The designs are either pebble bed reactors (PBR) or prismatic block reactors (PMR). The VHTR typically uses a graphite moderator with a helium coolant [70]. The VHTR can be designed with passive safety features [69], enabling the reactor to automatically shut down and reduce nuclear reactions if there is a problem.

A *sodium-cooled fast reactor* (SFR) is a fast neutron reactor that uses liquid sodium metal as the reactor coolant in a closed coolant system [69]. A fast neutron reactor uses a fast neutron spectrum, which means that the neutrons can react in a fission process without having to be slowed down with a moderator, as is done in other reactor designs [68]. This process is less efficient when uranium is used as the fuel and, therefore, fast reactors normally use plutonium as the fuel [71]. Using molten (or liquid) metal as the coolant with a solid core has the advantage of creating substantial thermal inertia against overheating should coolant flow be restricted or lost [72]. The reactor can also be used as a breeder reactor, to regenerate the fuel, reducing the need for new fuel and the problems associated with spent fuel [73]. These reactors are safer than current designs for two reasons. First, the reactor can be operated close to atmospheric pressures because the boiling point of sodium is higher than the operating temperature of the reactor. Second, molten salts cannot produce hydrogen, which is combustible [74]. The major drawback to the SFR is that sodium is highly reactive with air and water and any contact between them risks both the creation of toxic sodium oxide, along with possible explosions or sodium fires [69].

A *supercritical water-cooled reactor* (SCWR) is a high-temperature, high-pressure, fast reactor, cooled with supercritical water [75]. A supercritical fluid exists when the fluid is at a pressure and temperature above the critical point, so that there is no longer any distinctive liquid and gas phases, but the pressure is too low to force the substance into solid state. Operating above the critical pressure means that the coolant does not go through a phase change between liquid and gas; therefore, there is no need for many of the components needed to deal with the phase change, such as recirculation and jet pumps, steam generators, steam separators, and pressurizers [76]. Thus, SCWR plants are also considerably simpler mechanically. They also require a relatively smaller containment than current boiling water or light water reactors. SCWRs have much higher thermal efficiency at approximately 45 percent over current light water reactors, which have about 33 percent thermal efficiency. This makes them suitable for applications such as the production of hydrogen [76].

A *gas-cooled fast reactor (GFR)* is a variant on a sodium-cooled reactor [21]. It is a fast spectrum reactor that uses helium as a gaseous coolant, though CO_2 and steam have also been proposed [77]. GFR is viewed as an intermediary to the deployment of other gas-cooled thermal reactors, which makes the design work easier as it draws upon existing research and designs [78]. Like other fast reactors, the GFR is designed to use only spent fuel, relying on depleted or natural uranium to seed the reactions, which will then be regenerative reducing both the fuel used and the waste produced [79]. One of the advantages of helium is that it is not corrosive, making the system more sustainable [21]. However, the helium must be maintained at a high pressure and an appropriate pressure system has yet to be designed [74].

A *lead-cooled fast reactor* (LFR) uses molten lead (Pb) or lead–bismuth eutectic (LBE) as the coolant. Molten lead has a low melting point and high boiling point. Thus, it quickly solidifies in the event of a leak, supporting passive safety [80]. Lead and LBE do not react with water and air, which eliminates the need for an intermediary coolant system [81]. One of the drawbacks of using lead is that it is highly corrosive and requires highly corrosion-resistant components [21]. Several countries have worked on developing lead-cooled and LBE reactors, including the Soviet Union, Japan, the United States, and China [80].

Molten salt reactors (MSR) use molten salt (either fluoride or chloride) as both the base for the fuel mixture and the coolant [21]. The MSR is the most radical departure from Gen III designs. MSRs require the development of specialized materials and additional servicing of the graphite core during the reactor's operating life [21]. As with some of the other Gen IV designs, MSRs are designed to use spent fuel from other reactors as a source of fuel, reducing the need for new fuel and nuclear waste [68].

These designs are in various stages of research and development. To bring them to commercial energy production, significant investments are still needed. Some advancements have been made toward developing Gen IV reactors, particularly VHTR and SFR [50]. Other potential reactor designs are under development or investigation [82]. The next section discusses some of the investments that are being made in Gen IV reactors around the world.

10.3.1   Investments in the Development of Advanced Nuclear Reactors

Nuclear power has been used globally since the 1960s in both civilian and military applications. The United States and Russia both have long-standing nuclear power programs. Both are making public investments in the development of Gen III and Gen IV reactors, as is China. China has the stated goal of being a major exporter of nuclear reactors and technologies [83]. Organizations that develop technologies that dominate a market are often difficult to displace [84]. If either China or Russia come to dominate the market for advanced nuclear reactors, it will have serious implications for the current balance of power around the world. If either country is able to create a standardized advanced nuclear reactor with relatively certain costs for construction and operation, it would revolutionize the global market for nuclear energy. Countries with relatively low energy costs will be able to offer their goods and services much less expensively, giving them significant competitive advantages.

10.3.1.1   The United States

The United States has been making investments in advanced nuclear reactor research and development through the Department of Energy's Office of Nuclear Energy (US DOE ONE). The US DOE ONE is responsible for nuclear energy innovations, including developing new technologies and supporting the improvement of reactors. They are also charged with developing sustainable fuel cycles [85].[8]

For the fiscal year 2017, the US DOE ONE was allocated $994 million for its activities, including $90 million for small modular reactor (SMR) development project with NuScale; $109 million for new reactor concepts; $250 million for fuel cycle research and development; $90 million for nuclear energy enabling technologies; $365 million for the Idaho National Laboratory; and $5 million for the International Nuclear Energy Coordination program. The budget eliminated $5 million for the integrated university program and the STEP R&D, but significantly increased funding for SMR licensing technical support [86].

By 2020, the budget for the US DOE ONE had risen to $1.5 billion, including $230 million for the Advanced Reactor Demonstration Program [87]. The DOE has announced that it intends to build prototypes for two advanced nuclear reactors in the next seven years using public–private partnerships [88]. In addition to domestic investments, the United States has partnered with Canada, France, Japan, and the UK to jointly fund R&D on Gen IV reactors, agreeing to share any technical information gained from the project [50].

10.3.1.2   Russia

Russia has continued to be actively involved in the development and construction of nuclear reactors. Russia plans to replace all its current nuclear power plants with new ones. This requires commissioning a new nuclear power plant approximately every year until 2035 [89]. Russia also has an active export program. Rosatom, the state-owned company responsible for civilian nuclear energy, is currently constructing or operating reactors around the world, including Ukraine, China, Iran, India, Belarus, and Bangladesh, with additional reactors ordered by Turkey, Finland, Armenia, Egypt, Vietnam, Hungary, Slovenia, and Jordan [30].

Rosatom is now a major global nuclear manufacturer. It is building 28 of the 68 nuclear reactors currently under construction globally [90]. Part of Russia's success is in providing flexible financial arrangements for its customers [91]. Russia supports a model of build–own–operate (BOO), which allows customers to avoid the risks of nuclear power construction costs and overruns [92]. This is possible because the Russian government is actively involved in supporting nuclear exports [93]. The Ministry of Foreign Affairs promotes Russian nuclear technologies and President Vladimir Putin has directly engaged in some of the negotiations. Profits from Russia's fossil fuel industry have been used to support the nuclear industry, in an attempt to diversify Russia's energy industry and provide future economic stability [92].

The safety of Russian technology remains open to questions, given its history. However, Russia is vocal about being conscious of its safety record and in working to demonstrate the safety of its reactors [94]. They assert that Chernobyl made them conscious of safety in a way that was not possible before the accident and, thus, their reactors are safe because of Chernobyl, with more safety features built into their newer reactors [90].

The World Nuclear Association (WNA) reports on the nuclear industry and development in each country with a civilian nuclear program. The WNA reports that Russia is now working on developing advanced nuclear reactors [89]. They have been developing several research and commercial reactors with various designs. Their BN-600 fast neutron reactor has been in operation since 1980. Russia has developed the BREST reactor, a lead-cooled fast reactor [50]. In 2010, the Russian government approved a program designed to develop commercial fast reactors. Russia began construction on a multiloop research reactor in 2015. The BN-800 fast reactor has been operating since 2016 [83]. Rosatom plans to have fast reactors with a closed fuel cycle by the mid-2030s. The CEO of Rosatom stated that their goal is to make themselves the global leader in nuclear power construction and operation [30].

10.3.1.3 China

In the 1980s, China relied on technology transfer and foreign direct investments (FDI) to support their industries [95]. In the late 1990s and early 2000s, Chinese firms transitioned to a focus on indigenous innovation [96]. China is currently committed to developing its domestic reliance on nuclear power and in becoming a nuclear exporter. China has developed the domestic capacity to design and construct nuclear reactors, so that it is now largely self-sufficient in nuclear engineering [51].

The Chinese have been making significant investments in nuclear technologies. The Chinese government intends to become global exporters of nuclear reactors [51]. The WNA reports that China is investing billions of dollars into the development of their nuclear industry, including aggressively designing and building new reactors, with 17 currently completed, 30 under construction, and another 45–50 proposed and under review [51].

The Chinese are expected to surpass the United States in installed generating capacity by 2030. They have committed hundreds of millions of dollars to developing new reactors, including $350 million to a molten salt reactor and $476 million to a high-temperature-gas-cooled reactor [51]. The Chinese have invested political capital into developing their nuclear technologies. For instance, the molten salt reactor project was originally led by Jiang Mianheng, son of Jiang Zemin, the former President of the People's Republic of China and Secretary General of the Communist Party, which indicates significant political commitment to this project [97].

Sustained competitive economic advantage comes from leading in technological innovation and development, rather than simply following others. Though the Chinese obtained the original design for molten salt reactors from the experimental reactor that the United States built and operated at the Oakridge National Laboratories in the late 1960s and early 1970s, they are now investing heavily in building a working prototype, with the goal of producing a commercial reactor in the next 10–15 years [98,99]. China is constructing a prototype solid-fuel thorium molten salt reactor (TMSR) at the TMSR Research Center at the Shanghai Institute of Nuclear Applied Physics [50]. They have now invested $3.3 billion in the development of a molten salt reactor [100], and they are continuing to invest around $300 million per year [83]. They are the only country that is actually constructing a MSR [97].

China has developed a high-temperature gas reactor (HTGR), the HTR-10, completing cold functional tests in late 2020 [101]. They began operating a 10 MW helium high-temperature, gas-cooled (pebble fuel) test reactor in 2003. They recently completed a 200 MW prototype [102]. They started a 65 MW sodium-cooled fast reactor (SFR) in 2010, with a 600 MW commercial reactor expected to begin operations in 2023. China plans to build a 1,000 MW supercritical water-cooled reactor by 2022–2025 [103].

China has several stated goals in the development of their nuclear fleet [51]. The first is that pressurized light water reactors will be the main type of reactor, but they will diversify beyond this technology. The second is that nuclear fuel assemblies and equipment will be primarily designed and manufactured domestically. That is, their aim is to have an indigenous nuclear industry. Beyond this, China has ambitions to become an exporter of nuclear technologies and to leapfrog others in this area. The government of China listed nuclear power as one of its 16 science and technology priorities [104]. The first Chinese Hualong One reactor went into production in early 2021 with 90 percent domestically produced components. It is a domestically designed Gen III pressurized light water reactor [105]. China has stated that it is aiming to be a global exporter of nuclear reactors [83].

China is aggressively pursuing its nuclear strategy. China's construction of nuclear reactors represents about one-third of all global nuclear construction [106]. Thus, China is helping to offset the decline in global nuclear power [107].

10.3.1.4   Other Public Investments in Advanced Nuclear Reactors

Other countries are also investing in new reactors. In 2010, the European Commission announced the European Sustainable Nuclear Industrial Initiative (ESNII) [70]. The initiative was designed to support the development and prototyping of three Generation IV reactors: €5 billion ($5.7 billion) for a 500 MW sodium-cooled fast reactor (SFR) to be built in France starting in 2020; €1.96 billion ($2.2 billion) for a 75 MW lead-cooled fast reactor (LFR) to be built in Eastern Europe starting in 2020; and €1.2 billion ($1.4 billion) for a 300 MW gas-cooled fast reactor (LFR) to be built in Romania beginning in 2020 [70].

For the past few years, India has been actively working on developing a reactor to use its reserves of thorium as fuel. It currently has a small fast breeder reactor, and it is constructing a larger one. India is also working on the development of a molten salt breeder reactor [108]. The Indian government plans to build six more fast reactors, with the goal of creating thorium-based fast reactors in approximately 20 years [108].

These projects indicate that investments are being made in nuclear reactor innovations by governments around the world. Since developing Gen IV reactors are long-term undertakings, investments must be made decades in advance of actual deployment. As the next section shows, private investors want to develop these technologies, but generally they only pursue these investments in conjunction with government or philanthropic support.

10.3.2 Private Sector Investments in the Development of Advanced Nuclear Reactors

Private sector companies are also working on the development of advanced nuclear reactors. For example, TerraPower was founded by a consortium of investors led by Microsoft founder Bill Gates to develop new nuclear technologies. They are working with the Chinese National Nuclear Corporation to develop a traveling wave reactor. They have also been working on developing a molten salt reactor. In 2016, they were awarded $40 million from the US Department of Energy to research, design, and test a molten chloride fast reactor and another $80 million in 2019 to demonstrate their reactor and integrated energy system [109]. Transatomic Power was founded in 2011 by two graduates from the Massachusetts Institute of Technology (MIT) to develop a molten salt reactor [110]. UPower Technologies was founded by three MIT engineers in 2013 to develop an experimental breeder reactor [111]. Terrestrial Energy, founded in 2013 in Canada, is working on an integral molten salt reactor [112]. The Canadian Nuclear Safety Commission (CNSC) has agreed to a pre-licensing review of their technology [113]. Elysium Industries was founded in 2015 in Canada with the goal of developing a molten chloride salt fast reactor (MCSFR) [114]. Moltex, based in Britain, is working on a stable salt molten salt reactor [115]. Flibe Energy was founded in 2011 to develop a liquid fluoride thorium reactor (LFTR) [116]. These private sector efforts show that there are many companies that are investing in advanced nuclear reactors as part of the future of energy production.

These investments indicate that many countries expect that nuclear energy production is going to continue. However, the design and development of an advanced nuclear reactor is only part of the process for commercializing and deploying the reactor. The energy market, financing options, and governance structure are also important factors in creating a healthy nuclear power industry.

It is unclear whether the environment needed for widespread deployment of advanced nuclear power in the United States is there. The commercial nuclear industry in the United States has struggled for decades. There are now 96 reactors, down from 113 in the early 1990s [88]. Of the 53 nuclear reactors currently under construction globally, only two located in the United States [117]. Thus, the development of the technology is not necessarily going to lead to its adoption domestically. It remains to be determined whether US manufacturers will be leaders in these advanced nuclear technologies or whether other countries' reactor designs will dominate the global market. There are also questions about whether the current regulatory process in the United States is hindering the commercial development and deployment of nuclear reactors, putting US-based designs at a competitive disadvantage.

10.4 REGULATIONS

In the United States, the NRC has been charged with ensuring the safe development and use of civilian nuclear power in the country. With the ongoing wave of international advancements and global sales, the Chinese and Russian nuclear regulations and safety standards are of interest and concern. While the NRC once claimed to be the global standard for nuclear energy regulation and safety, this position is now in doubt, as more Russian and Chinese reactors are sold and deployed globally.

Originally, the US Congress established the Atomic Energy Commission (AEC) to regulate and advocate for nuclear power. However, this led to questions about regulatory capture and conflict of interests associated with regulating and advocating for a technology. Thus, the AEC was broken up in 1974, with the NRC taking on regulatory responsibilities and the research division being segregated and later absorbed into the Department of Energy [118]. The NRC is expressly forbidden from advocating for any specific technology or design. In fulfilling an explicit mandate of safety and effective oversight, the NRC has avoided the regulatory capture that can occur in other industries. The NRC has regulated the nuclear industry to ensure that safe use of nuclear power technologies is the primary objective [119].

The NRC has three major activities that it regulates [120]:

1. Nuclear reactor and facility construction, operation, and decommissioning.
2. Nuclear material possession, use, processing, exporting, importing, and transportation.
3. Spent fuel or waste disposal siting, designing, constructing, operating, and decommissioning.

A license is required from the NRC for any company that wants to build or operate a nuclear power plant in the United States. To get a license, an applicant (either an individual or organization) first submits an application to the NRC. The application is evaluated on both the technical merits (including safety) and environmental impacts. This licensing process is governed by US Code of Regulations 10 CFR[9] Part 50 [121].

The NRC is a fee-based regulator. That is, applicants and/or regulatees are responsible for paying the full cost of licensing. This cost is substantial. An applicant must pay for both the preparation of the application and the processing fee once the application is submitted [122]. The average 2021 cost per professional staff-hour was $288 [123]. Since the amount of time for the review by the NRC is not known in advance, the cost of licensing is uncertain ex-ante. That is, the total cost is only known once the obligation to pay for it has been undertaken.

There is also an annual regulatory fee charged by the NRC on each operating reactor [124]. In 2021, this fee was $5,050,000. This includes the annual fee for the reactor, a spent fuel storage and decommissioning fee, and additional associated charges. There are different fees for reactors being decommissioned or nonoperating with spent fuel. Research reactors are charged an annual fee of $78,800 [125].

The process of designing, building, and licensing a nuclear reactor has numerous steps, governed by 10 CFR Part 52. In general, the process requires:

1. developing a viable reactor design;
2. testing and evaluating the design with a prototype or research reactor;
3. developing a commercial design;
4. preparing the application (one to two years);
5. going through a regulatory review of the commercial design that certifies the design (five to six years);
6. selecting, preparing, and getting approval for a specific site (one to two years);
7. constructing the commercial reactor, including an environmental and site review (five to six years); and
8. obtaining an operating license for the reactor.

Steps 4 through 8 can take approximately 12–15 years: two years to prepare the application [126]; five to six years to approve a new reactor design, including time for public review, once the NRC has the complete application [127]; four years for the regulatory review itself [126]; and another five to six years to construct the power plant, assuming that there are no undue delays [128]. This is when the commercial reactor design is complete. Developing a new reactor design can add many years to the process, particularly when there is no previous experience with an actual (similar) reactor.

Figure 10.1 illustrates the development, deployment, and licensing process under the NRC. The first stages of design research and development, prototype testing and evaluation, and commercial design are under the purview of the US Department of Energy. That process is more organic, allowing for some overlapping and feedback. The stages of Regulatory Review and Licensing are under the NRC. This is a prescriptive process. It takes a minimum of 25 years for a new design to get to market. Throughout the duration of the process, the market opportunity and need, as well as the supporting private sector investments, human capital, the knowledge base, and public support for nuclear energy, all need to remain in place. A wall exists between the Department and Energy and the NRC. The DOE and the NRC have distinct mandates; therefore, systemic coordination and consistent goals are nonexistent. The NRC is not viewed as a partner in innovation. Instead, it considers itself an independent agency without technical bias or interest [129].

Whether these lengths of time are problematic for innovation is debatable. One of the primary functions of a regulatory body is to develop the rules and regulations for the activities that it is overseeing. Regulators and policymakers have to balance public protection and safety with commercial and economic considerations. Nuclear

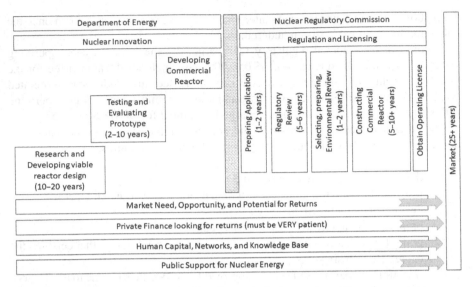

FIGURE 10.1 New nuclear reactor development, deployment, and licensing process.

regulations have to protect public interests and enable investments when they are in the public interest [122]. Unlike many industries that have many new entrants and new venture failures, the nuclear industry cannot risk these failures. Companies have to spend millions of dollars and many years developing technologies. To ensure ongoing public support, the regulatory process in the nuclear industry must be open and transparent, which necessarily makes it a slower, more deliberate process [130]. Many of the features of advanced nuclear reactors are theoretical. Their designs are unproven. Therefore, regulators need to carefully review and assess these new designs [122]. This necessarily makes the approval process slow, but it does not necessarily mean that the process is cumbersome. On the other hand, private sector organizations and finance must consider the expected returns versus the investment time and challenges. Investing in the development of new nuclear reactor designs and construction requires significant human capital and patience from investors.

In a normal innovation process, the different phases of development, testing, and evaluation, and deployment are not entirely sequential nor easily delineated. Often, there is significant overlap and feedback between these phases. However, within the nuclear industry, the process tends to be more linear than in other industries. No company can build a nuclear reactor in the United States until it has been licensed by the NRC. Companies are naturally reluctant to go through the expense of developing and completing a design for a commercial reactor unless they can be assured that their design has a reasonable expectation of regulatory approval. In addition, potential licensees must bear all of the regulatory and licensing costs. Thus, even beginning to have any preliminary discussions about reactor designs can be prohibitively expensive.

To determine whether regulations are hindering innovation, important questions need answers. The first is whether the NRC is engaging in discussions of advanced reactor development or just awaiting applications. If it is the latter, then the follow-up question is whether the regulatory structure prevents the NRC from engaging in these discussions. To answer these questions, we analyze the way that advanced reactors are discussed by the NRC.

10.5 NRC DISCUSSIONS OF GEN IV REACTORS

One of the ways to assess the progress of new nuclear designs toward commercialization and the role of the NRC in the innovation process is to investigate the conversations between the NRC and its stakeholders. Presumably, as advanced technologies get closer to commercial licensing and operation, the volume of communications and discussions about these technologies should increase. In addition, the communications around safety and operations should increase, while the communications around design and development should correspondingly decrease (since the closer the reactors get to commercial operation, the more fixed the design should be, and the less development of the design should be going on).

In order to analyze discussion within the NRC, it is necessary to use an appropriate methodology. This chapter employs text data analytics, grounded in corpus and computational linguistics. At the start of this process, documents are gathered and then transformed into an analyzable corpus that can be investigated using linguistic tools and techniques [131]. The first step in analyzing the corpus is to develop linguistic markers around the concept under investigation [132]. The set of search terms defining the concept under investigation is called a marker set. Marker sets are developed through an iterative process of investigating the results within the corpus to ensure that the returns are genuinely related to the concept under investigation.

Text has three forms of complexity [133]. The first is the technical complexity of the corpus (i.e., the difficulties of gathering and managing the data). The second is the complexity of language itself. The third type of complexity is in the concept under examination.

Marker sets were developed to investigate how each of the stages of innovation for Gen IV reactors are discussed by communications to and from the NRC.

The search terms used for Gen IV reactors are:

brayton cycle turbine*; generation iv; generation four; gen iv /3 reactor*; gen four /3 reactor*; gas cooled /3 reactor*; gas turbine modular helium; helium /3 reactor*; lead cooled /3 reactor*; liquid metal /3 reactor*; molten salt /3 reactor*; next generation /3 nuclear; next generation /3 reactor*; fluoride salt cooled; pebble bed /3 reactor*; sodium /3 reactor*; supercritical water cooled; super critical water cooled; high temp* /3 reactor; and vhtr.

The data for this study comes from the NRC's AMPS data. That is, the publicly available documents for the NRC. The available documents between 2001 and 2015 were gathered and converted into an analyzable corpus. The distribution of these documents is shown in Table 10.2. There are a total of 901,103 documents used in the analysis, with an average of 60,075 documents each year. In the corpus, there is a total of 5.411 billion words, with an average of 360.8 million each year.

TABLE 10.2
NRC Corpus Descriptive Statistics and Gen IV Occurrences

Year	Number of files	Number of words	Gen IV occurrences per million tokens
2001	42,605	245,248,744	8.30
2002	43,480	259,849,377	14.74
2003	54,129	323,706,196	9.77
2004	46,527	279,383,980	10.58
2005	48,358	256,265,543	9.03
2006	47,307	273,857,113	13.49
2007	58,650	329,250,439	5.74
2008	65,395	372,865,261	9.40
2009	63,866	438,590,196	5.45
2010	65,799	432,058,385	6.14
2011	84,810	441,133,690	6.13
2012	65,490	550,954,815	5.46
2013	99,500	503,442,812	3.04
2014	51,314	375,924,344	4.84
2015	63,900	328,896,631	2.37
Average	60,075	360,761,835	7.63
Standard deviation	15,730	95,060,151	3.56
Total	901,130	5,411,427,526	114.48

For the searches, an occurrence of the search terms in the marker set is recorded each time the word appears. The total number of occurrences indicates the number of times that the words occurred in the whole corpus. This can be difficult to interpret, since the number of occurrences is also dependent on the size of the corpus. To account for this variation, the number of occurrences is converted to a standardized number by dividing the total by the number of million words in the corpus. This yields the occurrences per million words, which makes it possible to compare the occurrences per year or even between corpora. Table 10.2 shows the occurrences per million words for Gen IV reactors. Figure 10.2 shows the occurrences per million words visually.

Gen IV reactors are currently conceptual and beyond the mandate of the NRC. Therefore, it is expected that there will not be a significant number of occurrences. What is surprising is the trend line. If there had been some progression in the development of Gen IV reactors, it would be expected that the discussions would increase as potential applicants engage more with the NRC on their reactor designs and future licensing requirements. Instead of the expected upward trend, there is a downward one.

The results of the analysis show that there was little progress toward Gen IV licensing between 2001 and 2015. In fact, there was actually less communication and correspondence at the end of the period then the beginning.

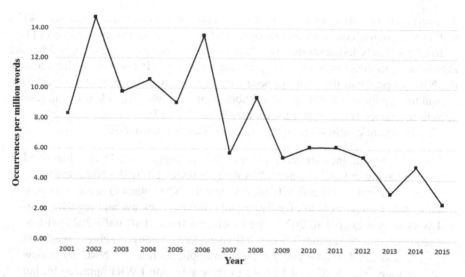

FIGURE 10.2 Gen IV occurrences in the NRC corpus.

10.6 CONCLUSIONS

Our analysis strongly suggests that the NRC is not actively engaging in discussions on advanced nuclear reactors. Though there are innovative activities and private sector engagements in the nuclear industry in the United States—with funding from the Department of Energy for private sector companies working on nuclear science and advanced reactor designs—these are not yet filtering into discussions involving the NRC in any substantive way. It is not that the NRC does not engage with the DOE on other issues, they do. It is just that with respect to new reactor designs, the NRC remains silent.

The NRC is also keenly aware of these concerns and its place in the nuclear industry and the innovation process. In December 2016, the NRC issued the "NRC Vision and Strategy: Safely Achieving Effective and Efficient Non-Light Water Reactor Mission Readiness, Document ML16356A670" [129]. The document states:

> The NRC is fully capable of reviewing and reaching a safety, security, or environmental finding on a non-LWR design if an application were to be submitted today. However, the agency has also acknowledged the potential inefficiencies for non-LWR applications ... that are reviewed against existing LWR criteria, using LWR-based processes, and licensed through the use of regulatory exemptions and imposition of new requirements where design-specific review, analysis, and additional engineering judgement may be required.

(p. 8)

The report goes on to acknowledge that the NRC has a significant amount of work to do in order to be prepared for non-LWR applications. They will need to establish "processes, procedures, and internal guidance ... for non-LWRs" [129, p. 8]. Their near-term strategy (0–5 years) is to acquire non-LWR knowledge, technical skills, technical capacity, computer technologies and tools to perform these reviews [129, p. 16]. This clearly indicates that they do not currently have these resources. They are also looking to develop regulatory guidance for non-LWR reviews, which requires the NRC to establish the "criteria necessary to reach a safety, security, or environmental finding for non-LWR applicant submissions" [129p. 16], "identify and resolve current regulatory framework gaps for non-LWRs" (p. 17).

The document continues to outline the timelines for non-LWR:

> The NRC has aligned its readiness activities to support the DOE's identified goal of having at least two non-LWR designs reviewed by the NRC and ready for construction by the early 2030s. As such, the NRC plans to achieve its strategic goal of readiness to effectively and efficiently review and regulate non-LWRs by not later than 2025. The timeframe from 2016 until 2025 will be used to execute the agency's non-LWR vision and strategy to achieve readiness … [A] non-LWR vendor could present an application to the NRC for review at any time. The NRC will be able to review [a non-LWR] application, but early applications will not benefit from the efficiencies gained as the non-LWR vision and strategies are implemented.
>
> [129, p. 20]

In other words, the NRC is willing to review applications before 2025, but does not yet have the capability or capacity to do so. This means that any applicant would face substantial delays as the NRC tried to catch up to the technology in the application and simultaneously develop the framework and procedures for evaluating the application. At this point, however, there is no expectation that Gen IV reactors will be developed and deployed before the 2030s.

This contrasts with the timelines for Gen IV reactors in Russia and China. Russia is looking to have commercialized Gen IV reactors by 2020–2030 [30], while China is targeting to have a commercial design in operation by the early 2020s [51].

The need for sustainable energy will continue, as fossil fuels become less desirable and available. A nuclear technology that is truly sustainable, safe, and able to address the spent fuel issue would be a disruption to energy markets if the technology also proves to be economically competitive. Significant disruptions to a market require multiple technologies coming together to interact and combine [134]. Thus, for the kind of disruptive change to energy markets that would be needed for genuine sustainability, there need to be multiple innovations and innovators.

As the empirical study suggests, the regulatory/financial bottleneck that prevails in the United States, due to the current operational procedures of the NRC, is a significant obstacle to groundbreaking nuclear innovation coming from the developed world. Currently, it appears most likely that innovative nuclear reactor designs are going to be deployed and made commercially available first in China and possibly Russia. Both the nations of the developing world, and well as the developed economies of the "Western" or "Global North" will then become dependent on them for their designs and technical expertise in advanced nuclear power generation. This raises questions regarding the different safety cultures of different nations, as well as having broader geo-political implications. A more robust response to these issues, as well as reducing the impacts of rapid climate change, would be for both the United States, along with other advanced nations, to work toward alleviating the political, economic, and societal barriers to nuclear innovation. This would contribute to a robust global marketplace where various technologies can compete—rather than allowing the situation to emerge where a single authoritarian regime controls the technology, deployment, safety protocols, and profits from widely deploying advanced nuclear power plants.

A major element to supporting a more broadly based drive toward deployment of innovative nuclear power plants must be persuading the citizens of the developed nations that nuclear energy should be viewed as a beneficial component to building societies powered by clean energy. Some of the strongest proponents for clean energy are environmental activists, specifically climate activists. But many of these stakeholders are either skeptical or antagonistic toward nuclear energy. From almost the beginning of its use, environmentalists and opponents criticized nuclear power [cf. e.g., 135, 136]. Others criticized the nuclear industry, including its costs, and the politics surrounding it [cf. e.g., 137–139]. Some still argue that nuclear power is more environmentally harmful than fossil fuels [cf. e.g., 140]. Schellenberger asserts that the underlying reason for the animosity of the environmental left is that nuclear does not offer the same potential for fundamentally remaking society as a switch to purely renewable community-based grids. Many climate activists seem to have a blind spot for the potential ecological impact of massive deployments of solar and wind energy [141]. However, that position is debatable, and scientists and environmentalists are starting to question this position. NASA scientists pointed out that fossil fuels are far more harmful than nuclear power [142]. Other activists and environmentalists are reluctantly accepting the necessity of nuclear energy [143].

Those advocating for decisive action against Climate Change ought to strongly support the rapid expansion of innovative nuclear energy technologies, but, as of this writing, most are not. Many problems need to be addressed and advanced reactors are still in the future. Advocates for nuclear power are going to need to find more effective ways of arguing for the technology. For several decades, nuclear industry advocates and their political allies have been ineffective in arguing on behalf of the technology. One problem is that the industry needs to be forthright and transparent about problems, letting other concerned stakeholders openly voice their anxieties. Another problem is that technical arguments are not necessarily persuasive, because emotions are often much more powerful in human decision-making processes. So nuclear energy advocates will need to effectively marry positive emotional persuasion with technical arguments. This turnaround needs to come very quickly.

In his recent book, Bill Gates advocates for using both renewable energy *and* nuclear power. He sees that the climate disaster facing us is too significant to leave any clean energy source unused [144]. No single energy source or technological solution is going to eliminate the use of fossil fuels and still provide sufficient energy to maintain our lifestyles [145]. Supplementing renewable energy with fossil fuels is not a viable solution. Neither is continuing to promote nuclear energy without a sustainable and reasonable solution to the spent fuel problem. Renewable energy and nuclear power are not alternative paths to climate change mitigation. They are both key components to decarbonizing the global energy supply chain, while still meeting energy demands. However, unless advanced nuclear reactors are embraced as a serious alternative, the regulatory barriers to the design, development, and deployment of these reactors will make it difficult for nuclear power to be a viable complement for renewable energy or an alternative to fossil fuels.

NOTES

1 For example, the loss of wildlife habitat is one of the concerns around the creation and spread of new diseases, such as the novel coronavirus (i.e., COVID-19) [9].

2 All the reactor designs discussed in this chapter are fission reactors. Deriving usable energy from controlled fusion processes (combining two atoms of hydrogen to create one helium atom) is still at an experimental stage, with any potential commercial deployment estimated to be at least several decades in the future.

3 This is what happened with the Fukushima Daichii power plant in March 2011 [42] ANS, "Fukushima Daiichi: ANS Committee Report," American Nuclear Society, LaGrange Park, IL 2012.

4 Some cooling is still required when a reactor shuts down, but it is significantly less than when the reactor is operating and can normally be maintained with backup cooling systems.

5 Spent fuel is the more appropriate term for nuclear waste. Spent fuel is the portion of uranium that is no longer suitable for use in the current nuclear reactor designs. France reprocesses and reuses spent fuel [47], which significantly reduces its waste problem. However, the United States is restricted from reprocessing its spent fuel under restrictions put in place by President Jimmy Carter over security concerns about the proliferation of weapons-grade fissile material [46].

6 The AP1000 design was approved in 2005 by the NRC (NRC, 2017b). The design for the AP1000 began in the 1980s along with the smaller AP600 reactor [55].

7 Units 3 and 4 at the Vogtle plant in Georgia, owned by Southern Company and Units 2 and 3 at the Virgil C. Summer plant in South Carolina, owned by the South Carolina Electric and Gas Company.

8 Spent fuel and current nuclear waste management are handled by the Office of Environmental Management in the Department of Energy.

9 U.S. Code of Federal Regulation.

REFERENCES

[1] IEA. (2020, March 13, 2021). *Renewables 2020, Fuel Report—November 2020.* Available: www.iea.org/reports/renewables-2020

[2] U.S. Energy Information Administration. (2021, March 3, 2021). Renewables account for most new U.S. electricity generating capacity in 2021, U.S. Energy Information Administration, January 11, 2021. Available: www.eia.gov/todayinenergy/detail.php?id=46576

[3] O. Ellabban, H. Abu-Rub, and F. Blaabjerg, "Renewable energy resources: Current status, future prospects and their enabling technology," *Renewable and Sustainable Energy Reviews,* vol. 39, pp. 748–764, 2014.

[4] A. R. Dehghani-Sanij, E. Tharumalingam, M. B. Dusseault, and R. Fraser, "Study of energy storage systems and environmental challenges of batteries," *Renewable and Sustainable Energy Reviews,* vol. 104, pp. 192–208, 2019.

[5] IPCC. (2018, March 9, 2021). *Energy Supply, IPCC Report.* Lead Authors: Ralph E.H. Sims and Robert N. Schock. Available: www.ipcc.ch/site/assets/uploads/2018/02/ar4-wg3-chapter4-1.pdf

[6] D. Pimentel, M. Herz, M. Glickstein, M. Zimmerman, R. Allen, K. Becker, et al., "Renewable energy: Current and potential issues: Renewable energy technologies could, if developed and implemented, provide nearly 50% of US energy needs; this would require about 17% of US land resources," *BioScience,* vol. 52, pp. 1111–1120, 2002.

[7] J. Jenkins. (2015, March 14, 2021). How Much Land Does Solar, Wind and Nuclear Energy Require? *Energy Central,* June 25, 2015. Available: https://energycentral.com/c/ec/how-much-land-does-solar-wind-and-nuclear-energy-require

[8] U.S. Census. (2010, March 14, 2021). *State Area Measurements and Internal Point Coordinates.* Available: www.census.gov/geographies/reference-files/2010/geo/state-area.html

[9] E. Gosalvez. (2020, March 9, 2021). How Habitat Destruction Enables the Spread of Diseases Like COVID-19, *North Carolina State College of Natural Resource News,* April 22, 2020. Available: https://cnr.ncsu.edu/news/2020/04/habitat-destruction-covid19/

[10] M. Sellenberger. (2019, March 2, 2021). Why Renewables Can't Save the Climate, *Forbes,* September 4, 2019. Available: www.forbes.com/sites/michaelshellenberger/2019/09/04/why-renewables-cant-save-the-climate/?sh=537ad1363526

[11] D. Proctor. (2020, March 13, 2021). China Seeks Grid Parity for Renewable Energy, *Power Magazine,* August 3, 2020. Available: www.powermag.com/china-seeks-grid-parity-for-renewable-energy/

[12] D. B. Poneman. (2019, March 2, 2021). "We can't solve climate change without nuclear power: Renewable energy, carbon-capture technologies, efficiency measures, reforestation and other steps are important—but they won't get us there," *Scientific American,* May 24, 2019. Available: https://blogs.scientificamerican.com/observations/we-cant-solve-climate-change-without-nuclear-power/

[13] World Nuclear Association. (2020, February 20, 2021). *Nuclear Power in the World Today,* Updated November 2020. Available: www.world-nuclear.org/information-library/current-and-future-generation/nuclear-power-in-the-world-today.aspx

[14] U.S. Energy Information Administration. (2021, March 13, 2021). *Short-Term Energy Outlook,* March 9, 2021. Available: www.eia.gov/outlooks/steo/data.php

[15] U.S. Energy Information Administration. (2017, March 2, 2021). *Most U.S. Nuclear Power Plants Were Built between 1970 and 1990.* Available: www.eia.gov/todayinenergy/detail.php?id=30972

[16] J. G. Marques, "Evolution of nuclear fission reactors: Third generation and beyond," *Energy Conversion and Management,* vol. 51, pp. 1774–1780, 2010.

[17] World Nuclear Asssociation. (2020, February 21, 2021). *Advanced Nuclear Power Reactors.* Available: www.world-nuclear.org/info/Nuclear-Fuel-Cycle/Power-Reactors/Advanced-Nuclear-Power-Reactors/

[18] G. Locatelli, M. Mancini, and N. Todeschini, "Generation IV nuclear reactors: Current status and future prospects," *Energy Policy,* vol. 61, pp. 1503–1520, 2013.

[19] World Nuclear Association. (2021, February 21, 2021). *Plans for New Reactors Worldwide.* Available: www.world-nuclear.org/information-library/current-and-future-generation/plans-for-new-reactors-worldwide.aspx

[20] U.S. Energy Information Administration. (2016, May 3, 2017). Fort Calhoun becomes fifth U.S. nuclear plant to retire in past five years, October 31, 2016. Available: www.eia.gov/todayinenergy/detail.php?id=28572

[21]   T. Abram and S. Ion, "Generation-IV nuclear power: A review of the state of the science," *Energy Policy,* vol. 36, pp. 4323–4330, 2008.

[22]   I. N. Kessides, "The future of the nuclear industry reconsidered: Risks, uncertainties, and continued promise," *Energy Policy,* vol. 48, pp. 185–208, 2012.

[23]   J. A. Lake, R. G. Bennett, and J. F. Kotek. (2009, January 23, 2021). "Next generation nuclear power: New, safer and more economical nuclear reactors could not only satisfy many of our future energy needs but could combat global warming as well," *Scientific American,* January 26, 2009. Available: www.scientificamerican.com/article/next-generation-nuclear/

[24]   IAEA. (2009) *Issues to Improve the Prospects of Financing Nuclear Power Plants.* Vienna, Austria: International Atomic Energy Agency.

[25]   IAEA. (2008, January 22, 2021). *International Nuclear and Radiological Event Scale (INES).* Available: www-pub.iaea.org/MTCD/Publications/PDF/INES2013web.pdf

[26]   D. Smythe. (2011, April 10, 2018). "An objective nuclear accident magnitude scale for quantification of severe and catastrophic events," *Physics Today: Points of View, December 12, 2011.* Available: http://large.stanford.edu/courses/2017/ph241/corti2/docs/smythe.pdf

[27]   IAEA. (2008, April 10, 2018). *International Nuclear and Radiological Event Scale (INES) Explanation and Significant Events.* Available: www.google.ca/url?sa=t&rct=j&q=&esrc=s&source=web&cd=11&cad=rja&uact=8&ved=2ahUKEwj8grXFr-3dAhWi24MKHQHWAIkQFjAKegQIARAC&url=https%3A%2F%2Fwww.iaea.org%2Fsites%2Fdefault%2Ffiles%2Fines.pdf&usg=AOvVaw2vNcii_bgcO8Zp5daqGbUZ

[28]   J. S. Walker, *Three Mile Island: A Nuclear Crisis in Historical Perspective.* Los Angeles, CA: The University of California Press, 2004.

[29]   B.-A. Schuelke-Leech, "Socio-economic implications of nuclear power on rural communities," in *Our Energy Future: Socioeconomic Implications and Policy Options for Rural America,* D. Albreicht, Ed., New York: Taylor and Francis, 2014, pp. 83–101.

[30]   World Nuclear Association. (2021, March 12, 2021). *Safety of Nuclear Power Reactors.* Available: www.world-nuclear.org/information-library/safety-and-security/safety-of-plants/safety-of-nuclear-power-reactors.aspx

[31]   D. McCaffrey, *The Politics of Nuclear Power: A History of the Shoreham Nuclear Power Plant.* Boston, MA: Kluwer Academic Publishers, 1990.

[32]   C. Perrow, "Fukushima and the inevitability of accidents," *Bulletin of the Atomic Scientists,* vol. 67, pp. 44–52, 2011.

[33]   S. Dekker, *The Field Guide to Understanding "Human Error".* Burlington, VT: Ashgate Publishing, 2014.

[34]   A. Prasad, "Forward-looking improvements to licensing the next generation of nuclear reactors," *American University Business Law Review,* vol. 2, pp. 209–223, 2012.

[35]   U.S. Energy Information Administration. (2017, December 30, 2017). Cost and Performance Characteristics of New Generating Technologies, *Annual Energy Outlook 2017* Available: www.eia.gov/outlooks/aeo/assumptions/pdf/table_8.2.pdf

[36]   U.S. Energy Information Administration. (2016, December 30, 2017). Levelized Cost and Levelized Avoided Cost of New Generation Resources, *Annual Energy Outlook 2017.* Available: www.eia.gov/outlooks/aeo/pdf/electricity_generation.pdf

[37]   M. Schneider and A. Froggatt, *World Nuclear Industry Status Report.* Mycle Schneider Consulting Project, 2014.

[38]   M. Schneider and A. Froggatt, *The World Nuclear Industry Status Report 2017.* Paris: Mycle Schneider Consulting Project, 2017.

[39] M. Berthélemy and L. E. Rangel, "Nuclear reactors' construction costs: The role of lead-time standardization and technological progress," *Energy Policy,* vol. 82 (2015), pp. 118–130, 2015.

[40] V. H. M. Visschers and M. Siegrist, "How a nuclear power plant accident influences acceptance of nuclear power: Results of a longitudinal study before and after the Fukushima disaster," *Risk Analysis,* vol. 33, pp. 333–347, 2013.

[41] S. Stapczynski. (2017, January 15, 2018). *Next-Generation Nuclear Reactors Stalled by Costly Delays.* Available: www.bloomberg.com/news/articles/2017-02-02/ costly-delays-upset-reactor-renaissance-keeping-nuclear-at-bay

[42] ANS, "Fukushima Daiichi: ANS Committee Report," American Nuclear Society, LaGrange Park, IL2012.

[43] B. Wheeler. (2011, March 2, 2021). *Gen III Reactor Design.* Available: www.power-eng.com/nuclear/gen-iii-reactor-design/#gref

[44] U.S. Department of Energy. (2020, December 18, 2020). *Benefits of Small Modular Reactors (SMRs).* Available: www.energy.gov/ne/benefits-small-modular-reactors-smrs

[45] D. Levitan. (2020, March 2, 2021). "First U.S. small nuclear reactor design is approved: Concerns about costs and safety remain, however," *Scientific American,* September 9, 2020. Available: www.scientificamerican.com/article/first-u-s-small-nuclear-reactor-design-is-approved/

[46] J. Mahaffey, *Atomic Awakening: A New Look at the History and Future of Nuclear Power.* New York: Pegasus Books LLC, 2009.

[47] G. Hecht, *The Radiance of France: Nuclear Power and National Identity After World War II.* Cambridge, MA: The MIT Press, 1998.

[48] J. S. Walker, *The Road to Yucca Mountain: The Development of Radioactive Waste Policy in the United States.* Berkeley, CA: University of California Press, 2009.

[49] GIF. (2020, January 19, 2021). *Gen IV International Forum.* Available: www.gen-4.org/gif/jcms/c_9260/public

[50] World Nuclear Association. (2020, February 22, 2021). *Generation IV Nuclear Reactors.* Available: www.world-nuclear.org/information-library/nuclear-fuel-cycle/ nuclear-power-reactors/generation-iv-nuclear-reactors.aspx

[51] World Nuclear Association. (2021, February 22, 2021). *Nuclear Power in China.* Available: www.world-nuclear.org/information-library/country-profiles/countries-a-f/china-nuclear-power.aspx

[52] World Nuclear Association. (2021, February 28, 2021). *Nuclear Power in USA.* Available: www.world-nuclear.org/information-library/country-profiles/countries-t-z/usa-nuclear-power.aspx

[53] J. M. Pedraza, *Small Modular Reactors for Electricity Generation.* Cham, Switzerland: Springer, 2017.

[54] NRC. (2020, February 12, 2021). *Issued Design Certification—Advanced Passive 1000 (AP1000)* Available: www.nrc.gov/reactors/new-reactors/design-cert/ap1000.html

[55] J. J. Taylor, K. E. Stahlkopf, D. M. Noble, and G. J. Dau, "LWR development in the USA," *Nuclear Engineering and Design,* vol. 109, pp. 19–22, 1988.

[56] S. M. Goldberg and R. Rosner, *Nuclear Reactors: Generation to Generation.* Cambridge, MA: American Academy of Arts and Sciences, 2011.

[57] U.S. Energy Information Administration. (2018, October 11, 2018). What is the status of the U.S. nuclear industry? May 1, 2018. Available: www.eia.gov/energyexplained/ index.php?page=nuclear_use

[58] B. Plumer. (2017, March 2, 2021). U.S. Nuclear Comeback Stalls as Two Reactors Are Abandoned, *New York Times,* July 31, 2017. Available: www.nytimes.com/2017/ 07/31/climate/nuclear-power-project-canceled-in-south-carolina.html

[59]   A. Ward. (2017, January 15, 2018). Nuclear Plant Nears Completion after Huge Delays: Western Europe's First Atomic Power Station in 15 Years is Test of Areva Technology, *Financial Times,* May 18, 2017. Available: www.ft.com/content/ 36bee56a-3a01-11e7-821a-6027b8a20f23

[60]   O. Ydinvoimalaitos. (2020, March 13, 2021). *Kolmosreaktorin valmistuminen Eurajoen Olkiluodossa on viivästynyt useasti yli vuosikymmenen ajan.* Available: https://yle.fi/uutiset/3-11516011

[61]   K. Rapoza. (2017, January 15, 2018). A Bankruptcy That Wrecked Global Prospects of American Nuclear Energy, *Forbes,* April 13, 2017. Available: www.forbes.com/sites/ kenrapoza/2017/04/13/a-bankruptcy-that-wrecked-global-prospects-of-american-nuclear-energy/#40bdc56a17a1

[62]   World Nuclear Association. (2018, March 2, 2021). Westinghouse emerges from Chapter 11, August 2, 2018. Available: https://world-nuclear-news.org/Articles/ Westinghouse-sale-to-Brookfield-completed

[63]   M. K. Rowinski, T. J. White, and J. Zhao, "Small and medium sized reactors (SMR): A review of technology," *Renewable and Sustainable Energy Reviews,* vol. 44, pp. 643–656, 2015.

[64]   Z. Robock, "Economics solutions to nuclear energy's financial challenges," *Michigan Journal of Environmental and Administrative Law,* vol. 5, pp. 501–540, 2016.

[65]   N. Barkatullah and A. Ahmad, "Current status and emerging trends in financing nuclear power projects," *Energy Strategy Reviews,* vol. 18, pp. 127–140, 2017.

[66]   GIF. (2020, February 22, 2021). *Origins of the GIF.* Available: www.gen-4.org/gif/ jcms/c_9334/origins

[67]   Gen IV International Forum. (2020, February 2, 2021). *Technology Systems.* Available: www.gen-4.org/gif/jcms/c_40486/technology-systems

[68]   U.S. Department of Energy. (2018, March 1, 2021). *3 Advanced Reactor Systems to Watch by 2030,* March 7, 2018. Available: www.energy.gov/ne/articles/ 3-advanced-reactor-systems-watch-2030

[69]   GIF. (2018, March 7, 2021). *GIF R&D Outlook for Generation IV Nuclear Energy Systems: 2018 Update.* Available: www.gen-4.org/gif/jcms/c_108744/ gif-r-d-outlook-for-generation-iv-nuclear-energy-systems-2018-update?details=true

[70]   World Nuclear Association. (2020, February 22, 2021). *Generation IV Nuclear Reactors.* Available: www.world-nuclear.org/info/Nuclear-Fuel-Cycle/Power-Reactors/Generation-IV-Nuclear-Reactors/

[71]   World Nuclear Association. (2020, March 3, 2021). *Fast Neutron Reactors.* Available: www.world-nuclear.org/information-library/current-and-future-generation/ fast-neutron-reactors.aspx

[72]   C. Grandy. (2008, March 7, 2021). *US Department of Energy and Nuclear Regulatory Commission—Advanced Fuel Cycle Research and Development Seminar Series, Argonne National Laboratory, ANL-AFCI-238, August 2008.* Available: http:// large.stanford.edu/courses/2018/ph241/rojas1/docs/anl-afci-238.pdf

[73]   A. Rojas. (2013, March 7, 2021). Sodium-Cooled Fast Reactors as a Generation IV Nuclear Reactor, Submitted as coursework for PH241, Stanford University, Winter 2018, May 25, 2018. Available: http://large.stanford.edu/courses/2018/ph241/rojas1/

[74]   C. Jones. (2017, March 7, 2021). Aging Plant Modernization, Submitted as coursework for PH241, Stanford University, Winter 2018, February 23, 2017. Available: http://large.stanford.edu/courses/2017/ph241/jones-c1/

[75]   GIF. (2020, March 7, 2021). Supercritical-Water-Cooled Reactor (SCWR). Available: www.gen-4.org/gif/jcms/c_42151/supercritical-water-cooled-reactor-scwr

[76]	D. Danielyan. (2003, March 1, 2021). Supercritical-Water-Cooled Reactor System as One of the Most Promising Type of Generation IV Nuclear Reactor Systems. Available: http://citeseerx.ist.psu.edu/viewdoc/download?doi=10.1.1.613.8434&rep=rep1&type=pdf

[77]	P. Tsvetkov, "4 - Gas-cooled fast reactors," in *Handbook of Generation IV Nuclear Reactors*, I. L. Pioro, Ed. Sawston: Woodhead Publishing, 2016, pp. 91–96.

[78]	R. Stainsby, K. Peers, C. Mitchell, C. Poette, K. Mikityuk, and J. Somers, "Gas cooled fast reactor research in Europe," *Nuclear Engineering and Design,* vol. 241, pp. 3481–3489, 2011.

[79]	P. Anzieu, R. Stainsby, and K. Mikityuk. (2009, March 2, 2021). Gas-cooled fast reactor (GFR): Overview and perspectives. Available: www.gen-4.org/gif/upload/docs/application/pdf/2013-10/gifproceedingsweb.pdf#page=128

[80]	A. Alemberti, V. Smirnov, C. F. Smith, and M. Takahashi, "Overview of lead-cooled fast reactor activities," *Progress in Nuclear Energy,* vol. 77, pp. 300–307, 2014.

[81]	GIF. (2020, March 7, 2021). Lead-Cooled Fast Reactor (LFR). Available: www.gen-4.org/gif/jcms/c_42149/lead-cooled-fast-reactor-lfr

[82]	OECD Nuclear Energy Agency. (2014, May 3, 2017). Technology Roadmap Update for Generation IV Nuclear Energy Systems, OECD Nuclear Energy Agency for the Generation IV International Forum. Available: www.gen-4.org/gif/upload/docs/application/pdf/2014-03/gif-tru2014.pdf

[83]	D. Yurman. (2020, March 7, 2021). "A forecast for the future of Gen IV reactors ~ A 50/50 chance of success for three types," *Neutron Bytes,* February 7, 2020. Available: https://neutronbytes.com/2020/02/07/a-forecast-for-the-future-of-gen-iv-reactors-a-50-50-chance-of-success-for-at-least-three-types/

[84]	C. M. Christensen, *The Innovator's Dilemma: The Revolutionary Book that Will Change the Way You Do Business.* New York: HarperBusiness Essentials, 2003 [1997].

[85]	U.S. DOE Office of Nuclear Energy. (2017, January 2, 2018). *Office of Nuclear Energy.* Available: https://energy.gov/ne/about-us

[86]	U.S. DOE Office of Nuclear Energy. (2017, January 29, 2021). *Fiscal Year 2017 Budget Request for the Office of Nuclear Energy.* Available: https://energy.gov/sites/prod/files/2016/02/f29/2016.02.09%20-%20NE%20FY17%20Budget%20Request%20.pdf

[87]	U.S. DOE Office of Nuclear Energy. (2020, January 29, 2021). *Fiscal Year 2021 Budget Request for the Office of Nuclear Energy.* Available: www.energy.gov/ne/our-budget

[88]	A. Cho. (2020, March 2, 2021). U.S. Department of Energy rushes to build advanced new nuclear reactors, *AAAS Science,* May 20, 2020. Available: www.sciencemag.org/news/2020/05/us-department-energy-rushes-build-advanced-new-nuclear-reactors

[89]	World Nuclear Association. (2021, February 22, 2021). Nuclear Power in Russia. Available: www.world-nuclear.org/information-library/country-profiles/countries-o-s/russia-nuclear-power.aspx

[90]	A. de Carbonnel. (2013, December 29, 2017). Russian Nuclear Ambitions, Reuters, July 22, 2013. Available: www.reuters.com/article/russia-nuclear-rosatom/russian-nuclear-ambition-powers-building-at-home-and-abroad-idUSL5N0F90YK20130722

[91]	J. Conca. (2017, December 29, 2017). The Geopolitics of the Global Nuclear Landscape, *Forbes,* May 20, 2017. Available: www.forbes.com/sites/jamesconca/2017/05/20/the-geopolitics-of-the-global-nuclear-landscape/#247d3005f68c

[92] G. Evans. (2015, December 29, 2017). Russia: New Nuclear Tech Titan, *Power Technology*, October 21, 2015. Available: www.power-technology.com/features/featurerussia-new-nuclear-tech-titan-4647211/

[93] Reuters. (2016, December 29, 2017). Rosatom's Global Nuclear Ambition Cramped by Kremlin Politics, *Fortune*, June 6, 2016. Available: http://fortune.com/2016/06/26/rosatom-global-nuclear-kremlin/

[94] A. E. Kramer. (2011, October 20, 2018). Nuclear Industry in Russia Sells Safety, Taught by Chernobyl, *The New York Times,* March 22, 2011. Available: www.nytimes.com/2011/03/23/business/energy-environment/23chernobyl.html

[95] X. Fu, *China's Path to Innovation*. New York, NY: Cambridge University Press, 2016.

[96] X. Fu and Y. Gong, "Indigenous and foreign innovation efforts and drivers of technological upgrading: Evidence from China," *World Development,* vol. 39, pp. 1213–1225, 2011.

[97] J. Tennenbaum. (2020, March 4, 2021). Molten salt and Traveling Wave Nuclear Reactors: Two Advanced Nuclear Power Reactor Designs That Can Solve a Multitude of Problems, *Asia Times,* February 4, 2020. Available: https://asiatimes.com/2020/02/molten-salt-and-traveling-wave-nuclear-reactors/

[98] B. Wang. (2017, 10/18/2018). China spending US$3.3 billion on molten salt nuclear reactors for faster aircraft carriers and in flying drones, *NextBig Future*, December 6, 2017. Available: www.nextbigfuture.com/2017/12/china-spending-us3-3-billion-on-molten-salt-nuclear-reactors-for-faster-aircraft-carriers-and-in-flying-drones.html

[99] D. Yurman. (2008, October 18, 2018). "Recent developments in advanced reactors in China, Russia," *Neutron Bytes*, January 7, 2018. Available: https://neutronbytes.com/2018/01/07/recent-developments-in-advanced-reactors-in-china-russia/

[100] B. Wang. (2018, March 4, 2021). China has multi-billion projects developing liquid and solid fuel molten salt reactors, *NextBig Future*, August 28, 2018. Available: www.nextbigfuture.com/2018/08/china-has-multi-billion-projects-developing-liquid-and-solid-fuel-molten-salt-reactors.html

[101] World Nuclear Association. (2020, March 4, 2021). Cold tests completed at first HTR-PM reactor, October 20, 2020. Available: www.world-nuclear-news.org/Articles/Cold-tests-completed-at-first-HTR-PM-reactor

[102] A. L. Harvey. (2017, January 15, 2018). China Advances HTGR Technology. Available: www.powermag.com/china-advances-htgr-technology/

[103] D. Martin. (2014, October 15, 2015). China's Next Generation Nuclear Ambitions, November 25, 2014. Available: www.the-weinberg-foundation.org/2014/11/25/chinas-next-generation-nuclear-ambitions/

[104] J. McDonald. (2016, May 4, 2017). China sets sights on new global export: Nuclear energy, August 24, 2016. Available: https://phys.org/news/2016-08-china-sights-global-export-nuclear.html

[105] A. Lee. (2021, February 28, 2021). China's Hualong One nuclear reactor goes into service, *South China Morning Post*, January 31, 2021. Available: www.scmp.com/news/china/science/article/3119959/chinas-hualong-one-nuclear-reactor-goes-service

[106] B. Spegele and Y. Saito. (2016, May 18, 2017). Going nuclear: A Quarter Century After China Fired Up Its First Nuclear Reactor, The Country Is Poised to Become the World's Biggest Producer of Nuclear Power. *Wall Street Journal*. Available: http://graphics.wsj.com/china-nuclear-plant/

[107] C. Liu. (2015, May 18, 2017). Build Up: The first two new nuclear reactors since the meltdowns at Fukushima received approval, *Scientific America,* March 11, 2015. Available: www.scientificamerican.com/article/china-restarts-nuclear-power-build-up/

[108] World Nuclear Association. (2021, February 22, 2021). Nuclear Power in India. Available: www.world-nuclear.org/information-library/country-profiles/countries-g-n/india.aspx

[109] TerraPower. (2020, January 2, 2021). TerraPower—A Nuclear Innovation Company. Available: www.terrapower.com/about/

[110] Transatomic Power. (2021, January 2, 2021). About Transatomic Power. Available: www.transatomicpower.com/press/

[111] J. B. Lassiter III, W. Sahlman, and L. Kind, UPower's Technologies Inc., Case 9-816-054. Boston, MA: Harvard Business School, 2017.

[112] Terrestial Energy. (2021, January 2, 2021). About Terrestrial Energy. Available: www.terrestrialenergy.com/about-us/

[113] S. McCarthy. (2017, January 13, 2018). Terrestrial Energy's Molten Salt Nuclear Reactor Approved by National Regulatory for Pre-licensing Reviews, Globe and Mail, November 8, 2017. Available: www.theglobeandmail.com/report-on-business/industry-news/energy-and-resources/terrestrial-energys-molten-salt-nuclear-reactor-approved-by-national-regulator/article36884953/

[114] Elysium Industries. (2017, January 2, 2018). About Elysium Industries. Available: www.elysiumindustries.com/home-1/

[115] Moltex Energy. (2017, January 2, 2018). About Moltex. Available: www.moltexenergy.com/aboutus/

[116] Flibe Energy. (2017, January 2, 2018). Company. Available: http://flibe-energy.com/company/

[117] M. Anderson. (2020, March 2, 2021). Slow, steady progress for two U.S. nuclear power projects, *IEEE Spectrum*, May 20, 2020. Available: https://spectrum.ieee.org/energy/nuclear/slow-steady-progress-for-two-us-nuclear-power-projects

[118] NRC. (2020, January 14, 2021). History of the Nuclear Regulatory Commission. Available: www.nrc.gov/about-nrc/history.html

[119] NRC. (2012, February 22, 2021). NRC—Independent Regulator of Nuclear Safety (NUREG/BR-0164, Revision 9). Available: www.nrc.gov/reading-rm/doc-collections/nuregs/brochures/br0164/

[120] NRC. (2020, February 22, 2021). Nuclear Regulatory Commission Licensing. Available: www.nrc.gov/about-nrc/regulatory/licensing.html

[121] NRC. (2020, February 22, 2021). Nuclear Regulatory Commission: New Reactors. Available: www.nrc.gov/reactors/new-reactors.html

[122] M. V. Ramana, L. B. Hopkins, and A. Glaser, "Licensing small modular reactors," *Energy,* vol. 61, pp. 555–564, 2013.

[123] NRC. (2020, January 15, 2021). Nuclear Regulatory Commission: CFR 170. Available: www.nrc.gov/reading-rm/doc-collections/cfr/part170/part170-0020.html

[124] NRC. (2020, January 14, 2021). Nuclear Regulatory Commission: 171.15 Annual fees: Reactor licenses and independent spent fuel storage licenses. Available: www.nrc.gov/reading-rm/doc-collections/cfr/part171/part171-0015.html

[125] NRC. (2020, March 14, 2021). U.S. Nuclear Regulatory Commission Fiscal Year 2021 Fees. Available: www.govinfo.gov/content/pkg/FR-2021-02-22/pdf/2021-03282.pdf

[126] NEI. (2015, October 15, 2015). Nuclear Energy Institute: Licensing New Nuclear Power Plants. Available: www.nei.org/Master-Document-Folder/Backgrounders/Fact-Sheets/Licensing-New-Nuclear-Power-Plants

[127] NRC. (2015, October 15, 2015). Nuclear Regulatory Commission: Frequently Asked Questions About License Applications for New Nuclear Power Reactors, NUREG/BR-0468. Available: www.nrc.gov/reading-rm/doc-collections/nuregs/brochures/br0468/br0468.pdf

[128] G. Goldfinger. (2015, October 15, 2015). Why does building a new nuclear plant take so long? Available: www.quora.com/Why-does-building-a-new-nuclear-plant-take-so-long

[129] NRC. (2016, January 14, 2018). NRC Vision and Strategy: Safely Achieving Effective and Efficient Non-Light Water Reactor Mission Readiness, ML16356A670, December 2016. Available: www.nrc.gov/docs/ML1635/ML16356A670.pdf

[130] T. Smith, "Nuclear licensing in the United States: enhancing public confidence in the regulatory process," *Journal of Risk Research,* vol. 18, pp. 1099–1112, 2015.

[131] C. M. Darwin, "Construction and Analysis of the University of Georgia Tobacco Documents Corpus," PhD Dissertation, The University of Georgia, Athens, GA, 2008.

[132] B.-A. Schuelke-Leech and B. Barry, "Philosophical and methodological foundations of text data analytics," in *Frontiers of Data Science*, M. Dehmer and F. Emmert-Streib, Eds., Boca Raton, FL: CRC, 2017, pp. 459–480.

[133] B.-A. Schuelke-Leech and B. Barry, "Text data analytics for innovation and entrepreneurship research," in *Complexity in Entrepreneurship, Innovation and Technology Research—Applications of Emergent and Neglected Methods*, A. Kurckertz and E. Berger, Eds., New York: Springer, 2016, pp. 459–480.

[134] B.-A. Schuelke-Leech, "A model for understanding the orders of magnitude of disruptive technologies," *Technological Forecasting and Social Change,* vol. 129, pp. 261–274, 2018.

[135] I. C. Bupp and J.-C. Derian, *Light Water: How the Nuclear Dream Dissolved.* New York: Basic Books, 1978.

[136] A. J. Surrey, "The future growth of nuclear power: Part 2. Choices and obstacles," *Energy Policy,* vol. 1, pp. 208–224, 1973.

[137] J. L. Campbell, *Collapse of an Industry: Nuclear Power and the Contradictions of U.S. Policy.* Ithaca, NY: Cornell University Press, 1988.

[138] R. J. Duffy, *Nuclear Politics in America: A History and Theory of Government Regulation.* Lawrence, KS: University Press of Kansas, 1997.

[139] D. Pope, *Nuclear Implosions: The Rise and Fall of the Washington Public Power Supply System.* New York: Cambridge University Press, 2008.

[140] T. Larsen and A. Graviz. (2006, February 18, 2021). 10 Reasons to oppose nuclear energy. *Green America.* Available: www.greenamerica.org/fight-dirty-energy/amazon-build-cleaner-cloud/10-reasons-oppose-nuclear-energy

[141] M. Shellenberger. (2019, December 17, 2020). The Real Reason They Hate Nuclear Is Because It Means We Don't Need Renewables, *Forbes*, February 14, 2019. Available: www.forbes.com/sites/michaelshellenberger/2019/02/14/the-real-reason-they-hate-nuclear-is-because-it-means-we-dont-need-renewables/?sh=6fdd834f128f

[142] P. Kharecha and J. Hanesen. (2013, February 17, 2021). Coal and gas are far more harmful than nuclear power, April 22, 2013. Available: https://climate.nasa.gov/news/903/coal-and-gas-are-far-more-harmful-than-nuclear-power/

[143] R. Harris. (2013, February 18, 2021). Environmentalists Split Over Need for Nuclear Power, National Public Radio, December 17, 2013. Available: www.npr.org/2013/12/17/251781788/environmentalists-split-over-need-for-nuclear-power

[144] B. Gates, *How to Avoid a Climate Disaster: The Solutions We Have and the Breakthroughs We Need.* New York: Random House, 2021.

[145] F. Forsythe. (2020, March 2, 2021). Renewables versus Nuclear: It's Not a Competition, *The Hill Times*, December 21, 2020. Available: www.hilltimes.com/2020/12/21/renewables-versus-nuclear-its-not-a-competition/276244

Index

Note: Page numbers in *italics* indicate a figure and page numbers in **bold** indicate a table on the corresponding page.

Printed in the United States
by Baker & Taylor Publisher Services